SOCIAL STRATIFICATION IN SCIENCE

Jonathan R. Cole
&
Stephen Cole

SOCIAL STRATIFICATION IN SCIENCE

The University of Chicago Press
Chicago & London

THE UNIVERSITY OF CHICAGO PRESS, CHICAGO 60637
THE UNIVERSITY OF CHICAGO PRESS, LTD., LONDON

Printed in the United States of America

International Standard Book Number: 0-226-11338-8
Library of Congress Catalog Card Number: 73-78166

JONATHAN R. COLE is associate professor of sociology at Barnard College and the Graduate Faculties of Columbia University

STEPHEN COLE is professor of sociology at the State University of New York at Stony Brook.

For Robert K. Merton

CONTENTS

PREFACE

In recent years social scientists have given increased attention to problems of inequality, justice, and discrimination in American society. The influence of ascribed statuses on the life chances of individuals is being studied in an effort to estimate the "fairness" of social institutions in rewarding talent. Social scientists are trying to determine the extent to which so-called "irrelevant" characteristics influence the ways in which people are judged by social institutions and eventually reach social positions in the hierarchies of income, prestige, and influence. Of the major institutions in American society, science has received perhaps the least systematic attention. Little is known about how scientists achieve positions of renown. This book examines several aspects of a single basic question: is the stratification of individuals in science based upon the quality of scientific performance, or does discrimination obtain in the processes of status attainment? A more technical way of putting the same question would be to ask whether universalistic and rational criteria predominate as the basis for recognition in the social system of science.

Our interest in the sociology of science was stimulated by participation in a graduate seminar offered by Robert K. Merton. Initially, we were interested in the social processes which influenced "the rise and fall" of academic departments. The need for an adequate measure of the quality of departments and of the work produced by resident scientists led to a year-long quest and resulted in our use of the *Science Citation Index*. We never finished our "rise and fall" study, being continually sidetracked toward larger issues that called for a set of relatively small-scale empirical studies of the stratification and reward systems of science. We are continuing these studies today. This book represents the results of our work as of June, 1972.

The sociology of science is a growing specialty. There are an increasing number of doctoral students and researchers in the United States who count it among their special research interests. There is a burgeoning of interest among historians and philosophers of science in sociological aspects of the development of science; and the specialty has taken hold in many European nations, where an increased amount of speculative and empirical research dealing with science as a social institution is being done. When we began our work some seven years ago, one could almost count on the fingers of one hand the number of sociologists of science in the United States. But if we have approached a "take-off" stage, we remain a relatively small, young, and youthful specialty.

When specialties are very young, the types of exploratory research that are fruitful are quite different from the types of research needed in more mature and highly codified specialties. Often the first mappings of the central or core problems require the adoption of what Max Delbruck has called "the principle of limited sloppiness." The process of exploring a new specialty often parallels the initial stages in the emergence of a new field: a whole range of big problems must be addressed by a few men. What Delbruck meant was that a science in its initial phases cannot make rapid progress without a degree of looseness in handling some definitional, conceptual, and measurement problems. Of course, researchers must be aware of the points at which they are being imprecise, and Delbruck's words should not be used as an apologia for muddled thinking. But sometimes it is necessary to use an imprecise measure rather than no measure at all. Consider only two reasons why first approximations are necessary in the early stages of the development of a field. First, since there are so many important substantive areas that have not been cultivated at all, the questions addressed often are large, and the data assembled to answer these questions are often less than ideal. Second, there is little other empirical research which has been done by which measures can be validated.

Current work in the sociology of science is charting unexplored areas. In itself this does not distinguish our specialty from many others in sociology. Since the results reported in this book represent initial empirical probes into the area of stratification in science, the reader should be aware of typical problems which result from "being first." Take, for instance, our use of citations as a measure of the quality of scientific work. Some people may see that as typifying the pitfalls of crass empiricism. To be sure, citation counts are not the ideal measure of the quality of research performance. Yet, a case can be made (and is, in chapter 2) that citations represent a good rough indicator of quality, better by far than anything else thus far developed in the sociology of science, and that the adoption of the measure has allowed us to address a set of questions which were central but which required some measure of the quality of work. The use of citations as an indicator of "quality," then, is a case in point for Delbruck's principle.

Our research methodology differs in at least two respects from that typically used in quantitative social research. Most quantitative studies are based on the analysis of data from one large sample. Here we report the data from at least twenty different small studies we have conducted. In doing the research, we would design a small study to test a set of limited and specific hypotheses. The results of the first study would suggest additional hypotheses which we then tested by additional limited studies. This procedure, which closely approximates that used in some of the natural sciences, has the advantage of enabling us to extend the scope of our understanding without depending upon ad hoc explanations. It has the disadvantage of making it more difficult for the reader to follow precisely how the results were generated. Since the properties of multiple samples may be difficult to keep in mind, we have included brief descriptions of the major samples in Appendix A.

The second way in which our methodology differs is in the eclecticism of our analytic techniques, which include tabular, correlation, regression, and path-analytic procedures. Different modes of analysis were employed because of their greater applicability to specific data sets or substantive questions. Whenever possible, more than one technique was used to analyze the same data set, although only one procedure is, of course, reported. In general, we have found our substantive conclusions to be supported by all forms of analysis. This should surprise no one, since essentially all of the procedures rest on similar logical foundations.

Though this book deals with general problems of stratification in science, it does not cover most scientific fields or settings; it concentrates on modern American science, predominantly physics. Clearly, in some respects physics as a scientific discipline is atypical. This fact calls for comparative analysis of the structures of different physical, biological, and social sciences to see whether some of our conclusions based on studies of physicists hold for other disciplines.

The book also has a very heavy bias toward the study of university-based science. Although a strong argument can be made that the most significant discoveries in pure science are made predominantly at university departments and affiliated laboratories, the majority of individuals who call themselves scientists do not work at universities. Processes of stratification within nonuniversity contexts await the attention of others and is clearly not the focus of our work.

Whenever one works over an extended period of time on a major research undertaking, there is the possibility that changes in the phenomena being studied will occur. To some extent that is true of our research. When we began, in 1967, the job market in physics was quite a bit more open than it currently is. Today, owing to the shortage of jobs, it may be difficult for even highly qualified scientists to get them. And it is possible that when supply greatly exceeds demand there is a reduction in the degree of universalism used in hiring. This possibility most certainly deserves attention. The shortage of jobs in physics makes some of the tentative policy suggestions made in chapter 7 even more salient. We believe that the conclusions reached in chapter 5 on sex discrimination in science are more valid today than they were a few years ago. Also, we are confident that different conclusions would have been reached had the study been done twenty-five years ago.

About most collaborative efforts there is raised the inevitable question about the creative role played in the production of ideas by the two or more collaborators. Although scientists may underplay the seriousness of this issue, we know that most of them are concerned, if not distressed, about the proper allocation of credit. This book has been a truly collaborative effort. The interaction and mutual intellectual stimulation has been invaluable; and frankly, it is often difficult, if not impossible, to determine the origins of many of the ideas in this book. The name-ordering, consequently, is simply a result of J preceding S in the alphabet.

ACKNOWLEDGMENTS

Patterns of intellectual influence are often difficult to determine. There have been a great number of social scientists who have helped to develop the ideas of this book. We would like to express our appreciation to those who have contributed most directly to our efforts.

Robert K. Merton, to whom we dedicate this book, is the founding father of the sociology of science and certainly the intellectual father of our own research in this area. This book certainly could not have been written were we not able to stand on Merton's quite substantial shoulders. Only those who have worked with him, and those who have firsthand knowledge of his extraordinary ability as a critic, can fully appreciate the impact of his advice on a sociological monograph.

Harriet A. Zuckerman, our other collaborator in the Columbia Program in the Sociology of Science, has done much research that parallels our own. We have benefited enormously from her research and her comments on ours. Another major intellectual influence has been the pathfinding work of the historian of science Derek J. de Solla Price. His now classic study, *Little Science, Big Science,* was the starting point for much of our own research. For criticism of an earlier draft of this book, we owe special thanks to Bernard Barber, Warren Hagstrom, Lowell Hargens, and Kenneth Bryson. Robert W. Hodge, Donald J. Treiman, Kermit Terrell, and John S. Reed offered numerous ideas and much technical assistance in the use of regression techniques and the development of the causal models in chapter 4. Hanan C. Selvin, Eugene Weinstein, and Judith Tanur provided much useful methodological advice. Lewis Coser, Rose Laub Coser, and Norman Storer made useful theoretical suggestions. Ann H. Cole, Lorraine Dietrich, and Naomi Gerstel provided essential computer skills. We thank Helen Astin for her generosity in allowing us to do secondary analysis of some of her data on women doctorates. The Office of Scientific Personnel, and especially Clarebeth Cunningham, was of great help in gathering the data reported in chapter 5. The errors that remain are, of course, our own.

The research was supported from its inception by several grants from the National Science Foundation to the Columbia Program in the Sociology of Science. Stephen Cole was a Ford Foundation Faculty Fellow during 1971–72, when the book was completed.

Finally, our wives, Joanna and Ann, must be given encomiums for persevering through the long gestation period of this book, and for their sheer tolerance of us during those many days when we were virtually incommunicado. We hope to reciprocate when their own books go through similar prenatal care.

1 THE SOCIOLOGY OF SCIENCE

Science has traditionally been seen as a lonely enterprise. The names of the greatest men of science—the Newtons and Einsteins—evoke images of isolated individuals grappling with momentous, mind-boggling problems. In part, this imagery has been conditioned by the words of the great men themselves. Newton, for instance, conveyed this sense of science, when he mused:

> I do not know what I may appear to the world; but to myself I seem to have been only like a boy playing on the seashore, and diverting myself in now and then finding a smoother pebble or a prettier shell than ordinary, whilst the great ocean of truth lay all undiscovered before me.[1]

The image of the Curies working in their cold Parisian flat also typifies the stereotype of the ambience attending great discoveries. Indeed, it was not long ago that even knowledgeable observers regarded the development of science as largely or wholly the result of discoveries made by geniuses in the isolation of their laboratories.[2] The conception of scientists working alone and interacting only with nature may have been held by some past historians of science, but it is generally recognized today that science in fact develops within a community of interacting scientists.[3]

Interest in the social organization of science and the ways in which this organization affects scientific development has been slow

1. Quoted in Robert K. Merton, "Priorities in Scientific Discovery: A Chapter in the Sociology of Science," *American Sociological Review* 22 (December, 1957): 646.

2. The position of the genius in the development of science has been extensively discussed. Among the older works, see Robert H. Murray, *Science and Scientists in the Nineteenth Century* (London: The Sheldon Press, 1925).

3. Among the earlier historians of science there were exceptions to this view of the process of scientific discovery. Two that were aware of social influences were George Sarton, *History of Science and the New Humanism* (New York: Henry Holt & Co., 1931), and Alfred North Whitehead, *Science and the Modern World* (New York: The Macmillan Co., 1948).

in coming. In fact, the sociology of science has a relatively short tradition. For the better part of this century, interpretations of the development of science have been left in the hands of historians. The differences in the approaches of the historian and the sociologist of science are by no means clear to most scholars or laymen. Few scholars, including most scientists and even some sociologists, know what the sociologist of science tries to do. Acquaintance with the perspective of the sociology of science is critical to an understanding of this book. Consequently, a brief description follows of the primary orientations of the specialty and how its perspective differs from that of the specialty of the history of science.

Even a brief scanning of the works of scholars interested in science will turn up a variety of approaches to understanding scientific development. Most scientists and, for that matter, educated laymen typically think of the growth and development of science as being the outcome of purely intellectual processes. Scientific advance is seen as a result of the cumulation and movement of scientific ideas. Science is its own master. If science flourished in seventeenth-century England, it was because British scientists simply had more creative ideas than their colleagues in other nations. If a particular scientific field declines in popularity, it is because the field has become entangled in some temporary Gordian knot. When some brilliant scientist can cut through this knot, the field will resume its former rate of growth. Although sociologists of science do not in any sense deny that intellectual factors play a crucial role in the development of science, they maintain that social factors are also significant.

In order to clarify the different perspectives of the sociologist of science and the intellectual historian we have developed a typology of influences on scientific development. We have identified two dimensions in the study of scientific advance. The first dimension is whether the influence is internal or external to the institution of science; the second dimension is whether the influence is of a social or intellectual nature. Cross-classifying these two variables yields the typology presented in figure 1. Each cell in figure 1 represents an "ideal type," and we should keep in mind that, in practice, scholars usually do not depend on a single type of influence in their analysis of science.

By intellectual influence on scientific development we mean the influence of ideas. The ideas influencing science may originate either inside or outside the institutional boundaries of science. Most scientists, traditional historians of science, and other believers in the "ascendency of the intellect" prefer Type I explanations of scientific development. To the advocates of this position, the history of science is the history of ideas and the intellectual bases of the development of those ideas. Emphasis is accordingly placed on the breakthroughs made by scientific geniuses or on the historical periods in which there was an efflorescence of ideas within a specific scientific discipline. The concern is generally, although not invariably, with a detailed discussion of the work of particular men like Newton, Halley, Hooke, Lavoisier, Cuvier, Darwin, Gibbs, or Einstein. The focus rests on the intellectual antecedents to particular discoveries—antecedents which are other scientific ideas.

There is, of course, a continuum of thought within each of these broader types. On the one hand, there are those historians who are so committed to the ascendency of intellect that they explicitly

Figure 1. Typology of influences on scientific development

	Source of Influence on Science	
	Internal to the Institution of Science	External to the Institution of Science
Type of Influence		
Intellectual	Type I	Type II
Social	Type III	Type IV

deny the influence of social factors on scientific development; on the other, there are those who see this particular perspective as most appealing, without denying the importance of social variables. The argument of the former position is clearly stated by Marie Boas Hall:

> It must not be forgotten that the scientific revolution was, in the last analysis, an intellectual revolution—a revolution in what men thought about and the way in which they did this thinking. And however much it may be true that every man is a product of his times and influenced by the economic and social developments of those times, new ideas must ultimately emerge out of the intellectual climate rather than the purely social climate, for every man must build on the ideas of his predecessors. . . . When all is said and done the scientific revolution was an

intellectual revolution, and its roots and causes must be sought in intellectual developments.[4]

Consider only a few outstanding representatives of this perspective. The historian of science A. Rupert Hall takes this stance in his book *Ballistics in the Seventeenth Century*.[5] Hall points out that we can only understand the development of ballistics if we are familiar with the ideas of "motion" of the seventeenth-century scientists. The growth of ballistics stemmed, Hall wrote, from then current scientific interests and ideas.[6]

Certainly the work of Alexandre Koyré, perhaps the most influential historian of science in the twentieth century, reflects an overwhelming concern with explaining scientific development through intellectual antecedents. In *Galilean Studies* Koyré discusses Galileo's ideas on moving bodies mainly in terms of the ideas of contemporaries and predecessors of Galileo.[7] Herbert Butterfield, in his classic work *The Origins of Modern Science,* holds fast to this same orientation. In introducing his book and his point of view, Butterfield takes his stand:

> It will be necessary rather to look for the lines of strategic change, and to put the microscope upon those moments that seem pivotal—trying, for example, to discover the particular intellectual knots that had to be untied at a given conjuncture. It will concern us particularly to take note of those cases in which men not only solved a problem but had to alter their mentality in the process, or at least discovered afterward that the solution involved a change in their mental approach.[8]

Among contemporary American historians of science, two of the first rank, Charles Gillispie and Thomas Kuhn, also embrace

4. Marie Boas Hall, *Nature and Nature's Laws: Documents of the Scientific Revolution* (New York: Harper & Row, 1970), pp. 15–16.

5. A. Rupert Hall, *Ballistics in the Seventeenth Century* (Cambridge: Cambridge University Press, 1952).

6. Ibid. In fact, Hall in rejecting the point of view of the sociological historian of science, concludes his book by noting: "Ballistics was a science necessarily created in the course of the scientific revolution. It flourished or languished as the aspects of physical science and mathematical technique upon which it depended prospered or were neglected, offered easy results or opposed difficult obstructions. As an element in continuous historical tradition of science the theory of projectiles found a place in the two greatest treatises of physical science, and as such it should be considered by the historian: as naturally offered by science in the course of its evolution, not wrenched from it by the strong hand of economic necessity" (p. 165).

7. See Alexandre Koyré, *Etudes Galiléennes* (Paris, 1939). For a brief discussion of Koyré's influence, see Marie Boas Hall, *Nature's Laws*.

8. Herbert Butterfield, *The Origins of Modern Science* (New York: The Free Press, 1957), p. 8.

this orientation in their work. Although both of these men clearly welcome the additional perspective of the sociologist of science, within their work they emphasize the historical chain of scientific ideas. Gillispie's imaginative work, *The Edge of Objectivity*, analyzes a series of intellectual conflicts and breakthroughs, indeed turning points, in the histories of several physical and biological sciences.[9] But the focus is definitely on intellectual antecedents, on intellectual controversies which inevitably had social overtones, and on the struggles within the sciences to move to a more "objective" analytical level.

Finally, consider Kuhn's *The Structure of Scientific Revolutions*, which has evoked widespread debate about the processes of scientific development among historians and philosophers of science.[10] Kuhn's treatment of scientific revolutions and his discussion of "paradigms" and "normal science" do indeed represent a new departure in the historical treatment of scientific development, in that Kuhn acknowledges the important role that social variables play in the evolution and transition from one scientific paradigm to another. In his explanation, Kuhn continues, however, to emphasize the influence of existing scientific ideas on emerging new ones. He implicitly suggests the importance of the values of scientists and of socialization processes. However, as Kuhn notes, his work still places primacy in "the revolutionary process by which an older theory is rejected and replaced by an incompatible new one,"[11] when the older theory fails "to meet challenges posed by logic, experiment, or observation."[12] The variables that are viewed by Kuhn as being most powerful in explaining scientific development are primarily intellectual, although his ideas and presuppositions are compatible with a sociological orientation. In fact, Kuhn has recently stated that an explanation of what is meant by scientific progress will have to be undertaken by social scientists.

> Already it should be clear that the explanation must, in the final analysis, be psychological or sociological. It must, that is, be a description of a value system, an ideology, together with an analy-

9. Charles Coulston Gillispie, *The Edge of Objectivity: An Essay in the History of Scientific Ideas* (Princeton, New Jersey: Princeton University Press, 1960).

10. Thomas S. Kuhn, *The Structure of Scientific Revolutions* (Chicago: University of Chicago Press, 1961).

11. Thomas S. Kuhn, "Logic of Discovery or Psychology of Research?" in Imré Lakatos & Alan Musgrave, eds., *Criticism and the Growth of Knowledge* (Cambridge: Cambridge University Press, 1970), p. 2.

12. Ibid., p. 2.

sis of the institutions through which the system is transmitted and enforced. Knowing what scientists value, we may hope to understand what problems they will undertake and what choices they will make in particular circumstances of conflict. I doubt that there is another sort of answer to be found.[13]

Each of the works alluded to here represent fundamental, indeed seminal, contributions to the study of science; they have contributed to the knowledge of the processes that attend the development of scientific idea systems. It would not be difficult to generate an extended list of historical studies that have adopted this orientation and have contributed vastly to our knowledge of the life and work of scientists. They represent, however, only one view, albeit the most highly developed one, of the study of science. Alternative perspectives that we will consider serve to increase our understanding of scientific development and need not necessarily undermine the perspective of the intellectual historian of science.

There is, of course, a difficulty in precisely locating the boundaries among the types of influence depicted in figure 1. It is not always clear whether an influence on science is internal or external to its institutional boundaries. Many intellectual historians have at the same time analyzed how science is influenced by past scientific ideas (Type I explanations) and by idea systems external to science (Type II explanations). One clear example is Gillispie's book *Genesis and Geology,* in which he simultaneously analyzes the role of religious beliefs and scientific ideas on the development of geology in the nineteenth century.[14] Many other scholars have examined the effects of religious and philosophical ideas on the development of scientific ideas. The need to integrate astronomical ideas with religious ideas can be seen, for instance, as an important influence on the work of Ptolemy. And, as Merton and others have pointed out, the philosophical ideas of Francis Bacon had great significance in the development of seventeenth-century science.[15] For the most part, however, works with this perspective have examined the ways in which belief systems external to science have acted as constraints on scientific growth. For example, historians have frequently tried to demonstrate the incompatibility between religious ideas and sci-

13. Ibid., p. 21.

14. Charles Coulston Gillispie, *Genesis and Geology: The Impact of Scientific Discoveries Upon Religious Belief in the Decades before Darwin* (New York: Harper & Row, 1951).

15. Robert K. Merton, "Singletons and Multiples in Scientific Discovery," *Proceedings of the American Philosophical Society* 105 (1961): 470–86.

ence, although scientists themselves have rarely seen the two systems as antithetical.

Social influences on science may also be divided into those that are internal or external to science. Type IV in figure 1 includes scholarship which emphasizes the effect on science of economic, political, religious, and other institutions. There is a long tradition of studies dealing with such external social influences. Marx and Engels viewed the aims and materials of science as largely dependent upon economic conditions, although they granted some autonomy to the conceptual content.

> Where would natural science be without industry and commerce? Even this "pure" natural science is provided with an aim, as with its materials, only through trade and industry, through the sensuous activity of men.[16]

Indeed Engels viewed the revival of science largely as a response to the needs of the expanding middle class.[17] In the end, Marx and Engels saw science as mainly determined by dynamic processes located within the economy. They held fast to the view that: "The mode of production of material life conditions the social, political and intellectual life process in general. It is not the consciousness of men that determines their being, but, on the contrary, their social being that determines their consciousness."[18]

During the 1930s a number of monographs produced by Marxist scholars focused on what the authors called "the social roots of science."[19] Hessen, Crowther, Bernal, among others, examined the development of English science from the seventeenth to the twentieth century.[20] In an article entitled "The Social and Economic

16. Karl Marx and Friedrich Engels, *The German Ideology* (New York: International Publishers, 1947), p. 36.

17. Friedrich Engels, *Socialism: Utopian and Scientific* (Chicago: C. H. Kerr, 1910), pp. 24–25.

18. See Karl Marx, *A Contribution to the Critique of Political Economy,* in *Marx, Engels Selected Works* (Moscow: Foreign Languages Publishing House, 1962), 1:363. For a discussion of the work of Marx and Engels in relation to the development of the sociology of knowledge, see Robert K. Merton, *Social Theory and Social Structure* (Glencoe, Illinois: The Free Press, 1957), chap. 12.

19. For a more extended treatment of these monographs, see Bernard Barber, *Science and the Social Order* (New York: Collier Books, 1962), pp. 51–92.

20. See B. Hessen, "The Social and Economic Roots of Newton's 'Principia,'" in *Science at the Cross Roads* (London: Kniga Ltd., 1931); James Gerald Crowther, *British Scientists of the Nineteenth Century* (London: K. Paul, Trench, Trubner & Co., 1935); J. G. Crowther, *The Social Relations of Science* (New York: The Macmillan Co., 1941); J. D. Bernal, *The Social Functions of Science* (New York: The Macmillan Co., 1939).

Roots of Newton's Principia," Hessen argues that the problems worked on by seventeenth-century scientists were determined by the technical demands of transport, means of communication, industry, and war. "In his [Newton's] mechanics he was able to solve that complex of physiotechnical problems which the rising bourgeoisie had set for decision."[21] Although their perspective led some of the more dogmatic Marxists to excessively narrow interpretations, their work served to heighten awareness of the social influences on scientific growth.

Robert K. Merton, in much of his early work on science, emphasized the influence of social structure on scientific development. He was particularly concerned with the functional interdependence between science and the economic, military, and religious institutions in society. His classic work, *Science, Technology and Society in Seventeenth Century England,* first published in 1938, brought an enormous quantity of empirical evidence to bear on the issue of whether the value orientation associated with Puritanism influenced the "scientific revolution."[22] Merton was also concerned with how the economic and military needs of England at the time influenced the foci of scientific attention. His work, since dubbed by A. R. Hall the "externalist" position toward scientific development, has over the years since its initial publication triggered considerable debate on the role that sociological influences play in scientific progress.[23] There were, of course, other renowned scholars in the 1920s and 1930s, like Dorothy Stimson, Martha Ornstein, and G. N. Clark, who adopted a perspective similar to Merton's.[24]

Explaining the development of science with an emphasis on external social variables has not been limited to historical case studies. Several works have examined the relevance of external social factors to the progress of contemporary American science.

21. Hessen, "Newton's 'Principia,' " p. 204.

22. Cf. Robert K. Merton, "Science, Technology and Society in Seventeenth Century England," *Osiris,* vol. 4, pt. 2 (1938) (reprint ed., New York: Howard Fertig, 1970).

23. The debate over Merton's position has been an extended one. It is still going on in the history of science journals. For a list of works that deal both directly and indirectly with the Merton perspective, see the preface to the reprint edition of Merton's book.

24. See Dorothy Stimson, "Puritanism and the New Philosophy in Seventeenth Century England," *Bulletin of the Institute of the History of Medicine,* 3 (1935); Martha Ornstein, *The Role of Scientific Societies in the Seventeenth Century* (Chicago: The University of Chicago Press, 1928); G. N. Clark, "Social and Economic Aspects of Science in the Age of Newton," *Economic History,* 3 (1937): 362–79.

Among these are work by Duprée, Kornhauser, Marcson, and Weinberg, who have analyzed the influence that government and industry have on science.[25] Since the government and the business communities are today the two most important sources of financial support for scientific research, there can be no denial of their impact in channelling scientific resources in one direction rather than another. The scientific community recognizes that the pace and direction of change in American science is strongly dependent upon social factors external to the scientific establishment itself.

Finally, consider work that adopts the angle of vision of Type III explanations. Researchers with this perspective see science as a social community. They examine the internal social organization of science for clues to its development, with emphasis on the social structure and value system of science itself. Historians of science have paid relatively little attention to how the social organization of science influences scientific work. When we read about seventeenth-century science, we learn all about the discoveries of Newton, Hooke, Wren, Harvey, Boyle, and others, but very little about what type of social structures were used to transmit ideas and to facilitate interactions among these scientists. Thus historians usually pay little attention to the role Henry Oldenburg played in institutionalizing science in the seventeenth century, even though he was instrumental in the formation of the Royal Society.[26]

A good example of the use of the Type III perspective may be found in a study by Joseph Ben-David and Awraham Zloczower.[27] They were interested in explaining why Germany became the center of scientific progress in the middle of the nineteenth century. Taking physiology as an example, they use histories of science to demonstrate that more physiology discoveries were made in Germany

25. See, among others, A. Hunter Duprée, *Science in the Federal Government: A History of Policies and Activities to 1940* (New York: Harper & Row, 1957); William Kornhauser, *Scientists in Industry: Conflict and Accommodation* (Berkeley: University of California Press, 1962); Simon Marcson, *The Scientist in Industry* (New York: Harper, 1960); Alvin Weinberg, "Criteria of Scientific Choice," *Minerva* 1 (Winter 1963): 159–71.

26. For a discussion of Oldenburg's role in seventeenth-century science, see Harriet A. Zuckerman and Robert K. Merton, "Patterns of Evaluation in Science: Institutionalisation, Structure and Functions of the Referee System," *Minerva* 9 (January 1971): 66–100. For an excellent analysis of the organization of French science in the seventeenth century see Roger Hahn, *The Anatomy of a Scientific Institution: The Paris Academy of Sciences, 1666–1803* (Berkeley, The University of California Press, 1971).

27. Joseph Ben-David and Awraham Zloczower, "Universities and Academic Systems in Modern Society," *European Journal of Sociology* 3 (1962):45–84.

than in any other country. They trace the prosperity of German science to the structure of the German university system. Competition among the twenty decentralized universities created an expanding opportunity structure in new scientific specialities. When one university created a new scientific chair, the others would follow. Bright young science students would go into the fields in which they saw expanding opportunities. Central to the Ben-David and Zloczower thesis is the hypothesis that in modern societies the prosperity of science depends upon the number and type of scientific jobs available; which is in turn dependent upon the social organization of science itself. In this type of analysis the researcher is not concerned with the origins of a particular discovery but with how social factors influence the rate of discovery in various social settings. This analysis is "internal" to the institution of science but social in nature.

The foci of attention in current research adopting the Type III perspective may be seen by a partial enumeration of topics which have been the subject of empirical investigations within the past few years. Scientific progress is in part dependent on effective communication of scientific ideas. Extensive research has been done on how the social structure of science and the location of scientists in that structure affect the flow of scientific information.[28] Most scientists are motivated to some extent by the desire for peer recognition. Thus sociologists have studied problems arising from competition and collaboration in science.[29] They have also done a substantial amount of research on the criteria by which rewards are distributed in the scientific community.[30] The behavior of

28. See, among others, Herbert Menzel, "Planned and Unplanned Scientific Communication," in Bernard Barber and Walter Hirsch, eds., *The Sociology of Science* (Glencoe, Ill.: The Free Press, 1962), pp. 417–41; Derek J. de Solla Price, "Networks of Scientific Papers," *Science* 149 (1965):510–15; Nicholas C. Mullins, "The Distribution of Social and Cultural Properties in Informal Communication Networks among Biological Scientists," *American Sociological Review* 33 (1968): 786–97; Robert K. Merton, "The Matthew Effect in Science," *Science* 159 (January 1968):56–63; Diana Crane, *Invisible Colleges* (Chicago: University of Chicago Press, 1972).

29. Harriet A. Zuckerman, *Scientific Elites: Studies of Nobel Laureates in the United States* (Chicago: The University of Chicago Press, forthcoming); Warren Hagstrom, *The Scientific Community* (New York: Basic Books, 1965); Merton, "Priorities in Scientific Discovery."

30. Diana Crane, "Scientists at Major and Minor Universities: A Study in Productivity and Recognition," *American Sociological Review* 30 (October 1965): 699–714; Bernard M. Meltzer, "The Productivity of Social Scientists," *American Journal of Sociology* 55 (July 1949): 25–29.

scientists, like that of other people, is governed by a set of norms and values. Scientists are expected to make the results of their research public, and secrecy is condemned. Several sociologists have analyzed this set of norms and the conditions which contribute to deviant behavior of scientists.[31] Other problems analyzed from this perspective are how productivity and quality of research are influenced by the social contexts in which scientists work;[32] the development of new fields of scientific work;[33] and the ethics of experimentation on human subjects.[34]

Are the intellectual and social interpretations of science in opposition? It is our view that social and intellectual influences, both internal and external, interact either to facilitate or impede scientific advance, and that therefore the role of all four types of influence should be studied. However, ideas and social structure may influence fundamentally different aspects of scientific development. Sociological variables clearly influence the types of problems scientists are interested in (the foci of scientific attention) and the rate of scientific advance; but is the specific content of scientific ideas influenced by sociological variables? Asking this question leads us to conclude that the differences between intellectual and social influences are analytic and sometimes difficult to separate empirically. Let us consider the intellectual influences on a scientist creating a new idea. When asked what influenced him, he might make reference to the work of several of his colleagues. But why did he think that such work was interesting, correct, or worth paying attention to? As Kuhn has gone the furthest in pointing out, the processes through which ideas come to be thought worthwhile are sociological.[35] Thus, there are many sociological variables mediating between the printed word and the reception and use of that word by its audience. The intellectual influences on a scientist are determined not only by the objective state of his field but by social processes influencing his subjective perception. It is here where

31. Warren Hagstrom, *The Scientific Community;* Merton, "Priorities in Scientific Discovery"; Barber, *Science and the Social Order;* Norman W. Storer, *The Social System of Science* (New York: Holt, Rinehart and Winston, 1966).

32. Crane, "Scientists at Universities."

33. Nicholas C. Mullins, "The Development of a Scientific Specialty," *Minerva* 10 (January 1972):51–82.

34. Bernard Barber, John J. Lally, Julia Laughlin Makaruska, Daniel Sullivan, *Research on Human Subjects: Problems of Social Control in Medical Experimentation* (New York: Russell Sage Foundation, 1973).

35. Kuhn, *Scientific Revolutions*.

sociological and intellectual variables intertwine and become difficult to separate.[36]

Perhaps the most important contribution of the sociology of science has been its challenge to the psychologistic view of scientific development. This view holds that science moves forward as a result of the idiosyncratic creativity of isolated geniuses. The sociological perspective sees scientific ideas as the creation of individuals working within a community. A crucial idea in the sociological perspective is the "ripeness of time" concept. According to this concept a discovery is made at a particular time because conditions are ripe for that discovery. Thus, if Newton, Darwin, or Einstein had not made their particular discoveries, other scientists would have, although there would be in some cases delays in scientific advance. The evidence that sociologists use to support this idea are independent multiple discoveries. Merton, who has dealt more extensively than anyone else with the importance of "multiples," has noted: "At the root of a sociological theory of the development of science is the strategic fact of the multiple and independent appearance of the same scientific discovery."[37] The list of great minds involved in multiple discoveries is virtually all-inclusive. In 1922 William Ogburn and Dorothy Thomas catalogued about 150 cases of multiples.[38] Almost all of the giants in the history of science, from Galileo, Newton, Faraday, Kelvin, Darwin, Descartes, Huyghens, Lister, to Freud and the great men of contemporary science, have been involved in disputes arising out of multiple discoveries.

The existence of independent multiple discoveries is sometimes taken to be evidence for the social as opposed to the intellectual interpretation of science.[39] However, the ripeness of time concept is a necessary tool for both the intellectual historian and the sociologist of science. A discovery can be made independently by two or more scientists for many reasons. The intellectual state of a field can lead many scientists to work on the same problem, or a particular social condition can lead many scientists to work on and

36. This analysis will be further developed in chapter 3.

37. Merton, "Singletons and Multiples," p. 475.

38. W. F. Ogburn and D. S. Thomas, "Are Inventions Inevitable?" *Political Science Quarterly* (March 1927), pp. 83–98.

39. A. Rupert Hall, "Merton Revisited, or Science and Society in the Seventeenth Century," *History of Science* 2 (1963):1–16. Hall attributes the ripeness of time concept to what he calls the "externalists" (scholars using social variables to explain scientific development) and then rejects it. However, as shown below, Hall uses the concept in his own work.

ultimately solve the same problem. As an example of how the intellectual historian uses the ripeness of time concept let us look at a quote from A. Rupert Hall's analysis of the scientific revolution. Hall tells us that the time was ripe for Newton's discoveries, and indeed that these discoveries were inevitable.

> How much the fruit of this [Newton's] originality owed to the ground from which it sprang, to Barrow, to Wallis Slusius, Kepler, Borelli, More, Boyle, Descartes and others whom he read in those early Cambridge years! How many, and how rich, were the threads which these giants led to the hand of so able a spinner of theorems, so close a weaver of theories! The genius that was frustrated by the dense mysteries of chemistry reaped a splendid harvest in the *riper* fields of mathematical science in which its fullest powers could be exercised. It is no detraction from Newton's originality to point out that all his discoveries were firmly rooted in the science of his time—that like a helmsman he was borne along by a stream—that each of them has a quality of *inevitability* in its contemporary context. Newton, in fact, won such immediate esteem because he saw clearly the things to which others were groping, because he was so fully in harmony with his age.[40]

Does the "ripeness of time" concept negate the importance of great scientists? Our estimation of Newton as a great scientific genius is in no way undermined by our knowledge that his discoveries were affected by all four types of influence under discussion. Although all of his fundamental discoveries would have been made by others had Newton not lived, this does not detract from his social and intellectual significance in the history of science.[41] A great man of science, sociologically defined, is one who makes discoveries that would require many other men to duplicate. He is the functional equivalent of many other scientists. Moreover, an additional function may be served by the great man. By virtue of his reputation, a great man might speed the acceptance of a new idea. An innovation made by an unknown scientist might go unnoticed, or diffuse very slowly into the mainstream of scientific knowledge. A scientific

40. A. Rupert Hall, *The Scientific Revolution 1500–1800* (London: Longman's Green and Co., 1954), p. 247 (emphasis added).

41. For a detailed discussion of the sociological meaning of scientific genius see Merton, "Singletons and Multiples"; idem, "Resistance to the Systematic Study of Multiple Discoveries in Science," *European Journal of Sociology,* 4 (1963): 237–82.

discovery by a great man might draw immediate attention in the scientific community.[42]

We may conclude this point by noting that whether the ripeness of time concept is used by the intellectual historian or by the sociologist of science it calls our attention to the fact that science is a communal social enterprise. To understand this enterprise we must look at its communal aspects as well as at the role played by creative geniuses.

This brief outline of the perspective of the sociology of science is far from complete. We have discussed only work treating scientific development as the dependent variable. An important, although infrequently studied, part of the sociology of science would be the analysis of how science influences society. Such work would treat science as an independent variable. Some of the more polemical historians of science have gone so far as to argue that, rather than society influencing science, science influences society and that this influence should be the exclusive concern of the sociologist of science.[43] We would only point out that the path of causality between science and society is not uni-directional. To say that science (for example, the scientific revolution) was influenced by society (for example, the Protestant Reformation), does not imply that science is any less important than other social institutions. Indeed it is possible for an effect to be a far more historically significant event than its cause.

Having very briefly outlined the general orientation of the sociology of science, let us move to the particular topics covered in this book. We are explicitly concerned with the social structure of the scientific community and how that system operates to determine patterns of scientific behavior. This work, therefore, will concentrate on Type III influences on scientific development. Specifically, we shall examine in detail the system of social stratification as it exists in contemporary American science. Our work does not attempt to look at a chain of scientific ideas, except as such a chain might be influenced by the location of scientists in the social hierarchy of science. Nor will we examine the influences of other social institutions on science. These important areas of inquiry we leave for the moment to others. We are interested in the internal social organization of science and its operation. In fact, our focus

42. See Merton, "The Matthew Effect."
43. Hall, "Merton Revisited."

is even more restricted. Most of the empirical sections of this work rely on data gathered from the community of academic physicists.

The stratification system of science is a strategic starting point for an inquiry into the social system of science. A substantial part of the efficient operation of science depends upon the way in which it allocates positions to individuals, divides up the rewards and prizes it offers for outstanding performance, and structures opportunities for those who hold extraordinary talent. In short, understanding the processes that determine the social inequalities within the scientific community will move us a long way toward comprehending the way science works as a social institution.

We must realize that stratification or social inequality in science is inextricably interwoven with most of the other subsystems operating in science. Take the relationship between the reward system and stratification. The process of evaluating role-performance is a necessary condition for the emergence of stratification. Indeed, as Barber points out: *"the product of the interaction of social differentiation and social evaluation . . . is social stratification."* [44] Evaluation, in its concrete form, is represented by the distribution of socially valued recognition and rewards. These rewards may take a number of shapes. In science, as in most other institutions, prestigious position, honorific awards, and peer recognition, as well as monetary rewards, combine to form an integrated reward structure. The pattern of stratification in science is determined in large measure by the way rewards are distributed among scientists and by the social mechanisms through which the reward system of science operates to identify excellence.[45] Reward systems that fail to honor excellence often produce social conditions leading to deviance.[46] And it is problematic whether science as well as other social systems does honor excellence wherever it is found. Consider only one form of adaptation that might result from an imperfectly operating reward system. A popular notion in contemporary academia is that a

44. See Bernard Barber, *Social Stratification: A Comparative Analysis of Structure and Process* (New York: Harcourt, Brace, and World, 1957), pp. 2–3, author's italics.

45. Robert K. Merton, "'Recognition' and 'Excellence': Instructive Ambiguities," in *Recognition of Excellence,* Adam Yarmolinsky, ed. (Glencoe, Illinois: The Free Press, 1960), pp. 297–328.

46. The model of sources of deviant behavior that Merton presented in "Social Structure and Anomie," in *Social Theory and Social Structure,* pp. 131–60, has been applied in some degree to structural strain in science in his paper, "Priorities in Scientific Discovery."

scientist must "publish or perish." Surely this image has some basis in reality. But the basic question is whether scientists are rewarded primarily for the sheer bulk of their publications or for the quality of what they publish. If sheer quantity is rewarded, the possibility increases of a sharp displacement of scientific goals. If the system recognized quantity of work alone, scientists who were capable of significant contributions would often alter their publication habits, rushing into print with little regard for the intellectual substance of their research. They would choose their research problems with an eye to quick and sure results rather than to solving the important and difficult intellectual problems of the discipline. A total disregard for the quality of work would surely hinder scientific advance. Such an extreme situation does not exist in modern science. Yet little is known about the level of effectiveness of the reward system of science. Further, little is known about the consequence of a malfunctioning reward system for stratification in science. One of the aims of this study is to take a step toward understanding that system.

Scientific advance is dependent on the efficient communication of ideas. Plainly, only those discoveries which come to be known can have an impact on the development of science. Only then do they become functionally relevant for the advance of science. Without efficient communication of ideas, rewards could not be allocated properly. Recognition is conditional upon the visibility of scientific work. Visibility is a function of effective communication. Only through an analysis of the scientific network of communication can we establish the "payoff" to scientists of being strategically located in that communication system. The effective communication of scientific ideas is a necessary condition for the rational operation of the reward process as well.

The communications system, then, is the nervous system of science—the system that receives and transmits stimuli to its various parts. Communications systems can, of course, operate at different levels of efficiency. Ideas could freely circulate among members of the same social stratum in the system, for example, but not circulate at all among strata. Closed circles involving communication between scientists located at the most distinguished centers of scientific research might exist, while scientists removed from these centers might have little knowledge of what is being done at the research front.

In this book, then, we shall address ourselves to a set of basic problems, which may conveniently be formulated as questions. First, what are the contours of stratification in the scientific community? Second, what social processes determine the location of individual scientists in the social hierarchy? Third, how do the reward and communications systems influence the assignment of individuals to social positions? Fourth, what are the consequences for individuals and the social system of differential position within the scientific hierarchy? Fifth, how does social status affect the manner in which a scientist's research is received by the community? In other words, does work of equal quality meet different kinds of receptions within the community depending on the eminence of the author of that work? Sixth, do scientists at all levels of the stratification system contribute through their research to scientific advance, that is, what are the sources of intellectual influence on the production of outstanding scientific work? Seventh, does science, more than other institutions, approach the ideal of a meritocracy, in which the operation of universalistic criteria largely determines the contours of stratification? It should be clear from these questions that this book will indeed focus on the internal social organization of science.

In concluding this introduction, an overview of the content of the book may serve to place individual chapters in perspective. Before tackling the substantive issues which form the core of this work, we must attend to a basic measurement problem. In the past, empirical studies in the sociology of science have floundered to some extent because of an inability to satisfactorily measure the quality of scientific output. Since virtually all studies of scientific advance depend upon assessing the calibre of research produced by a large group of scientists, a convenient and simple measure of the quality of output had to be found. Without such a measure it would be impossible to determine the independent influence of the quality of work and of social variables on the stratification process. We believe that substantial progress has been made towards solving this measurement problem. The emergence of the *Science Citation Index* provides for the first time a rough but useful indicator of the quality or impact of scientists' work. Since much of the empirical discussion that follows is dependent upon the use of citation counts as a measure of quality, chapter 2 presents an analysis of the adequacy of this measure.

In chapter 3 we move to a qualitative and theoretical discussion of stratification in science. Flesh is placed on the bare-boned questions that we have raised in this introduction. Not only shall we discuss the shape of stratification in science with particular reference to the normative structure, the reward system, and the role of authority in determining social inequalities, but we shall also consider why there is relatively little identifiable social conflict between social strata that receive such vastly different proportions of rewards. In short, why is there not more "class conflict" within the scientific community?

In chapter 4 we examine the process of assigning individuals to social positions within the stratification system. Specifically, we look at the extent to which scientists who have produced various types of research are rewarded by the system. We identify four patterns of research publication based upon the relationship between the quality and quantity of a scientist's lifework. Does the scientific community reward scientists for the sheer bulk of their work? Or, does the quality of publication, regardless of the number of papers produced, make for greater recognition through receipt of honorific awards, appointments to prestigious academic departments, and being widely known to the scientific community? In short, we illustrate with data the responses of the community of physicists to distinct patterns of research publication. Finally, we attempt to get behind the correlation that exists between the quality and quantity of output in order to see how the reward system operates to encourage creative scientists, and, correlatively, to divert the energies of less creative scientists.

In chapter 5 we continue the analysis of the process through which individuals are located in the stratification system. However, where in the previous chapter we concentrated on how evaluation of the scientist is effected by his location in the scientific social structure, here we concentrate on how nonscientific statuses influence evaluation. To what extent do functionally irrelevant statuses such as sex, race, and religion influence the distribution of rewards in science? We pay particular attention to the question of whether women are discriminated against in science.

At this point the focus of the work shifts. Thus far we looked at the determinants of stratification in science. We now begin to concentrate on the consequences for science and scientists of being located in one or another social stratum. In chapter 6 we consider several aspects of the communications process in science. We examine the general knowledge that physicists throughout the system have of research produced by different types of physicists. Are physicists located at less distinguished universities as familiar with work produced by men at Harvard, for example, as are the physicists at universities of the first rank? In this way we will be dealing with what William James has called "knowledge about" information, or nonspecific forms of knowledge. We go on to analyze the diffusion of information about the formal reward system. Is there differential knowledge of the formal reward structure among men variously located in the stratification system of science? Finally, we consider the detailed knowledge of communications that is necessary for actual use in a scientist's own research. We ask, what are the characteristics of scientists who make use of research produced by different types of scientists? This chapter is addressed to the following basic questions in communications research: how does social structural position influence the flow of information within the scientific community; are there limitations of any kind to the free flow of information from one part of the social system to another? In brief, we examine the level of efficiency of the communication system of science.

In chapter 7 we analyze data on the reception of scientific research. We concentrate on how work of equal quality, produced by men of unequal rank and eminence, is received by scientists throughout the community. Does the work of eminent men diffuse more quickly and widely because of their existing reputations and the previous honors bestowed upon them? Or, do stratification variables have a limited or inconsequential effect on the reception of new discoveries? A series of small studies tests these alternative hypotheses. If the reception of equal quality work depends upon the characteristics of the men who produce it, then what are the consequences for scientists and scientific disciplines of this abridgment of the value of universalism?

Chapter 8 examines another crucial question that can be asked of any stratification system. To what extent do the members of various strata contribute to the achievement of social goals? To what extent do scientists variously located in the stratification system contribute to scientific progress? It is a widely held belief that the work of eminent scientists depends upon the mass of work produced by their less eminent colleagues and that therefore the larger the number of scientists in a particular field the faster the rate of advance. We shall examine some empirical evidence which allows us to make a preliminary test of the validity of this proposition.

The final chapter summarizes the major findings but is primarily concerned with a discussion of the application of rational and universalistic criteria in evaluating scientific performance. In a speculative way we try to indicate the consequences for individuals of a system that approaches a meritocracy. The influence of accumulative advantage and self-fulfilling prophecies on status attainment is discussed. We also consider possible criticisms of our analysis of the stratification system. The question of how scientists might cope with relative "failure" within a universalistic social system is considered.

2 MEASURING THE QUALITY OF SCIENTIFIC RESEARCH

The problem of assessing the quality of scientific publications has long been a major impediment to progress in the sociology of science. Most researchers have typically paid homage to the belief that quantity of output is not the equivalent of quality and have then gone ahead to use publication counts anyway.[1] There seemed to be no practical way to measure the quality of large numbers of papers or the lifework of large numbers of scientists. The invention of the *Science Citation Index* (*SCI*) a few years ago provides a new tool which we believe yields a reliable and valid measure of the significance of individual scientists' contributions.[2] Starting in 1961, the SCI has listed all citations to scientific work appearing in a large number of journals. Thus, it is possible to count the number of citations made in 1961 and 1964–73 to a particular book, paper, or work of a given scientist. The number of citations is taken to represent the relative scientific significance or "quality" of papers.

There is some supporting evidence for this assumption and procedure. In a thorough study of measures of scientific output, Kenneth E. Clark asked a panel of experts in psychology to list the psychologists who had made the most significant contributions in their

1. See Derek Price, *Little Science, Big Science* (New York: Columbia University Press, 1963); Myron B. Coler, ed., *Essays on Creativity in the Sciences* (New York: New York University Press, 1963); Logan Wilson, *The Academic Man: A Study in the Sociology of a Profession* (reprint ed, New York: Octagon Press, 1964), p. 110; Diana Crane, "Scientists at Universities." In fact, researchers have had a great deal of difficulty in estimating the significance of even a small number of papers. Although a panel of judges is occasionally used, problems of standardization of evaluation criteria and the individual biases of the evaluators have frequently been encountered.

2. The *Science Citation Index* is compiled under the direction of Eugene Garfield. In 1961 the *Index* listed all citations made in 613 journals; 1962–63 have not been indexed. In 1964 700 journals were covered, and in 1965 1,147 journals. In 1971 more than 2,000 journals were indexed. Virtually all important journals in the natural sciences are included. Social science journals are being added to the *SCI*'s file. In 1971 ten sociology journals were added to the file.

fields.[3] He then investigated the correlation between the number of choices received by psychologists and other indices of eminence. The measure most highly correlated with number of choices was the number of journal citations to the man's work (r = .67). Clark concludes that the citation count is the best available indicator of the "worth" of research work by psychologists.

Consider another kind of validating evidence for this measure. Recipients of the Nobel Prize in the physical and biological sciences can be regarded, in the aggregate, as having contributed greatly to the advance of their fields, even though the great scarcity of the Nobel Prize means that there are probably other like-sized aggregates of eminent scientists who may have contributed as much. Nevertheless, the laureates as a group can be safely assumed to have made outstanding contributions. The average number of citations in the 1961 *SCI* to the lifework of Nobel laureates who won the prize in physics between 1955 and 1965 was 58, compared to an average of 5.5 citations to the work of other scientists cited in 1961. Only 1 percent of the quarter of a million scientists who appear in the 1961 *SCI* received 58 or more citations.[4]

We thought it possible that winning the prize might make a scientist more visible and lead to a greater number of post-prize citations than the quality of his work warranted. We therefore divided the laureates into two groups: those who won the prize five or fewer years before 1961 and those who won the prize after 1961. The pre-1961 laureates were cited an average of 42 times in the 1961 *SCI;* the post-1961 prizewinners an average of 62 times. Since the prospective laureates were more often cited than the actual laureates, we conclude that the larger number of citations primarily reflects the high quality of work rather than the visibility gained by winning the prize.[5] Here, we have used receipt of the Nobel Prize as an inde-

3. Kenneth E. Clark, *America's Psychologists: A Survey of a Growing Profession* (Washington, D.C.: American Psychological Association), chap. 3. For another study using number of citations as a measure of quality of scientific work, see Alan E. Bayer and John Folger, "Some Correlates of a Citation Measure of Productivity in Science," *Sociology of Education* 39 (Fall 1966):381–90.

4. Irving H. Sher and Eugene Garfield, "New Tools for Improving and Evaluating the Effectiveness of Research." Presented at the Second Conference on Research Program Effectiveness, 1965, Washington, D.C. We thank Dr. Sher for making available some unpublished data. By way of emphasizing the difference between the number of citations to the work of laureates and to the work of the average scientist, we would also point out that many scientists do not appear in the *SCI* and that the modal number of citations to the work of those who do appear is 1.

5. These statistics are based upon the work of twenty-four of the twenty-eight living laureates who won the prize in physics as of 1965. The four living laureates

pendent measure of the quality of a scientist's work. In the studies reported on in this book we have found measures of quality based upon citations to be highly correlated with other measures of eminence. For example, we found the quality of work of 120 university physicists to be correlated ($r = .57$) with the number of awards received (see chap. 4). These data offer further support for the use of number of citations as an indicator of the scientific significance of published work.

Citation counts enable us to distinguish the extent of contributions by various types of physicists. Consider, for example, the citations to one relatively productive and eminent group of scientists: members of university departments of physics. Among a representative sample of 1,308 academic physicists, only 2 percent had a total of 60 or more citations to their work in the 1961 *SCI*. Thus, only a small fraction of university physicists have received a number of citations equal to that of the average laureate in physics. Another 12 percent of the sample of 1,308 had between 15 and 59 citations; and the remaining 86 percent had fewer than 15 citations to their life's work. In short, very few physicists are heavily cited, and there are distinct differences in the number of citations to physicists whose quality of research has been validated by other measures of eminence.

THE CONCEPT OF QUALITY

Whether citation counts are a useful measure of the quality of scientific work depends, of course, on how quality is defined. Quality may be defined in two different ways. A traditional historian of science might apply a set of absolute criteria in assessing the quality of a scientific paper. Those papers which embody scientific truth and enable us to better understand empirical phenomena are high-quality papers. The fact that a particular set of works may be momentarily in fad or temporarily ignored tells us nothing about the quality of the work, if we use the absolute definition. Using this definition, the quality of work could only be measured in historical retrospect.

Another way to conceptualize quality is to use a social as opposed to an absolute definition. The social definition is built upon

---

who won the prize more than five years before 1961 were excluded so as not to introduce an age bias. Included in these computations are the non-American laureates; when they are excluded, the average number of citations to the work of American Nobelists is 68.

the philosophical view that there is no absolute truth, and that truth is socially determined.[6] Since what is believed to be true today may not be true tomorrow, few if any scientific discoveries will ever meet the absolute criteria. In the long run all discoveries will be seen as being in some fundamental aspect incorrect. Therefore we define high-quality work as that which is currently thought useful by one's colleagues. If scientists in their everyday behavior find a particular idea useful in their work, that idea is a valuable one, and we shall call it a high-quality idea. Throughout this work "quality" is used strictly in this social sense. We make no claims that work receiving a high number of citations is superior, using absolute criteria, to work receiving fewer citations. We leave the ultimate assessment of scientific work to future historians. Current citations are not a measure of the absolute quality of work, they are an adequate measure of the quality of work socially defined.

Although citations provide us with an improved method of assessing the quality or impact of a scientist's research, there are a number of definite problems in the use of this measure. Most of these problems are basically substantive as opposed to technical. We consider the substantive problems first.

ERRORS IN EVALUATION

There may be occasional "errors" in the evaluation of scientific work. The significance of work done by a scientist is not always recognized immediately. New ideas (especially ones which lead to changes in the basic scientific paradigms) are sometimes resisted or ignored.[7] Some great scientific innovators remain obscure in their own time only to be accorded posthumous recognition. Mendel is the classic example of a scientist whose work was unappreciated by his contemporaries but greatly honored by scientists of later generations. Using citations to measure quality, we must necessarily misclassify work which is currently being "resisted" or is being judged inadequately. Since only history will reveal which work is being resisted or misjudged, this flaw in our procedure is inevitable. The problem of resistance to significant contributions, however, may be less pressing in contemporary science than in the past. In a study of delayed recognition of scientific discoveries, we examined citation patterns to papers over time (see Chap. 7). For a sample of

6. Kuhn, *Scientific Revolutions.* For a further discussion of this point of view see chapter 3.

7. Bernard Barber, "Resistance by Scientists to Scientific Discovery," *Science* 134 (September 1961):596–602.

177 papers published in the *Physical Review* we found a strong correlation between the number of citations the papers received one year after publication and the number they received after three years (r = .72). When we examined citation patterns to papers published in different scientific fields between 1950 and 1961, we found a similarly high correlation between the number of citations received by the papers in the 1961 *SCI* and the 1966 *SCI*. Although an ideal study of delayed recognition of scientific discoveries would require citation data over a longer time period, these data at least suggest that the communication and evaluation systems of modern science work effectively enough so that there are relatively few cases of research unrecognized after publication but judged significant at some later date. There will, of course, always be some cases of error in the evaluation system. If these errors are randomly distributed, then using citations as a measure of quality will not effect substantive conclusions. As sociologists we are interested not in idiosyncratic occurrences but in socially influenced patterns.

CRITICAL CITATIONS

Citations may refer to papers that are being criticized and rejected rather than utilized. It is unlikely, however, that work which is valueless will be deemed significant enough to merit extensive criticism. If a paper presents an error which is important enough to elicit frequent criticism, the paper, though erroneous, is probably a significant contribution. The significance of a paper is not necessarily determined by its correctness. As we pointed out above, much of the work of the great historical figures of science was in some sense "wrong" or mistaken. It is unlikely that any work which is wrong without being a "fruitful error" will ever accumulate many citations. Suppose we had a total of one thousand citations to scientific work, and as many as one hundred of these were to work being criticized or rejected. The majority of these "critical" citations are dispersed among a large number of papers, such that most papers cited critically would receive no more than one or two citations. The same type of dispersion is found for "positive" citations. Let us say that one paper actually receives as many as twenty-five "critical" citations. We suggest that this piece of research which has stimulated wide criticism has, in fact, stimulated other research. Consequently, this research must be considered mistaken but significant; it must be seen as work which has had an impact on future scientific research.

TREATING ALL CITATIONS AS EQUAL UNITS

If each citation to a paper is given equal weight, errors in assessing the impact of research may follow. A paper that is cited widely by scientists of the first rank should not be equated with a paper cited predominantly by scientists who have made only minor discoveries. Since citations do not have equal meaning, can we treat them as equal, or should we consider classifying citations by the characteristics of the citer? In effect, would a citation count weighted for the quality of research produced by the citer be a superior measure of the quality of scientific papers?

We collected data which suggest an answer to this problem. We did a detailed study of the citers of a subsample of the 1,308 university physicists. For each of the 171 physicists in the subsample, we collected data on a random sample of his citers. This enabled us to classify each physicist by the characteristics of his citers. By giving each citer a score depending upon the number of citations his work had received,[8] we were able to measure the quality of work of each *subject* physicist by the quality of work of his citers. The correlation between this index and the total number of citations received by the subject physicist was $r = .40$. We believe that this correlation is not higher for two basic reasons. (1) Since the study was designed for substantive purposes rather than methodological ones, we took a maximum of fifteen citers for each physicist. Therefore, the proportion of the total number of citers that our sample represents varies from one physicist to another. (2) Since a disproportionate number of citations are made by a small group of scientists who publish heavily and who themselves are generally highly cited, the same men are likely to be found among the citers of high-quality work and low-quality work. Thus, 62 percent of the citers of relatively low-quality work and 70 percent of the citers of relatively high-quality work received ten or more citations to their own work. Although the two indices of quality—the total number of citations received and the index based upon the quality of the citers—are not highly correlated, they yield similar correlations with other variables. For example, rank of academic department is correlated $r = .19$ with a straight citation count and $r = .22$ with the

8. Scores were assigned in the following way. Citers whose work had received 100 or more citations in the 1965 *SCI* were given a score of 5; citers receiving 50–99 citations were given a score of 4; those receiving 25–49, a score of 3; 10–24, a score of 2; 1 to 9, a score of 1; and those physicists whose work was not cited received a score of zero. These scores were then totaled for all citers and each subject physicist was classified by the total index score.

index. We conclude that a citation index in which characteristics of the citing authors were considered would probably yield similar substantive conclusions as an index which treats all citations as equal in value.

QUANTITY AND QUALITY OF RESEARCH OUTPUT

The total number of citations to a scientist's work may in part be dependent on the quantity of his work. It is possible that scientists who publish a large number of papers, each of which receives only a few citations, might accumulate as many citations as those who have published only a few papers which are heavily cited. In a sample of 120 university physicists, we found a correlation of .60 between the total number of papers a physicist had published and the total number of citations listed after his name in *SCI*. However, the correlation between total number of papers and the total number of citations to the three most heavily cited contributions of the scientists (a measure which is unlikely to be an artifact of the quantity of publications) is .72. This is exactly the opposite of what we would find if the total number of citations were primarily a function of the total number of papers. We concluded that the total number of citations could serve as an adequate indicator of quality.

In several of our studies we have chosen to use more refined measures of citations because these measures seemed substantively more suitable. For example, in a paper analyzing the relationship between the quantity and quality of physicists' output, we use the number of citations to the three most heavily cited contributions by each physicist as a measure of quality, in order to eliminate any possible effect of sheer productivity on the total number of citations. Moreover, since a contribution in physics does not typically take the form of a single paper but is usually presented in a series of papers, we used citations to the year's output rather than the single paper as the unit of measurement.[9] A further word is necessary on the relationship between the quantity and quality of a sci-

9. Stephen Cole and Jonathan R. Cole, "Scientific Output and Recognition," *American Sociological Review* 32 (June 1967):377–90. The year is also an arbitrary unit since physicists do not, of course, arrange their related papers to fit the calendar year. For a more exacting procedure, it would be necessary to identify the series of papers representing an integrated contribution, a requirement extremely difficult to meet in dealing with large numbers of working scientists. Without such detailed information, it would seem preferable to use a period of time as a unit rather than single papers. See also the recent study of scientific productivity by Crane, "Scientists at Universities," which treats a series of four papers on the same topic as a "major" publication, and single papers as "minor" publications.

entist's research. Since in physics these two variables are highly correlated, we may conclude that where citation counts are not readily available (as in historical research) publication counts are roughly adequate indicators of the significance of a scientist's work. In recent research we have found that the quantity and quality of research output are highly correlated in other fields.[10] Where citation counts are available, they should be used as they have definite advantages over paper counts.

SIZE OF SCIENTIFIC FIELDS

Differences in the size of the various scientific disciplines pose another potential problem in the use of citations as a measure of quality. Of two scientists who have produced work of roughly equal quality, will the one coming from the larger field receive a higher number of citations? It is incorrect to directly compare the quality of work of scientists working in dissimilar areas. Indeed, how could we say an outstanding physicist is a "better" scientist than, for example, an outstanding biochemist? The only way to make such a comparison would be to evaluate the relative position of the scientists in their respective fields. Thus the work produced by physicists or biochemists in the top 1 percent of their respective fields, as measured by citations, would be seen as roughly equal. We recommend that whenever a study contains scientists from more than one field the citation data should be statistically standardized separately for each field and the standardized scores used in the analysis.[11]

Standardization is a straightforward way to handle differences in size among different fields. It is more difficult to use the technique when considering differences in size among specialties in the same field, because it is frequently very difficult to classify the specialty of a scientist, and the work of many scientists overlap several specialties. Will the number of citations a scientist receives be heavily influenced by the number of people working in his research area? If there are more publications in solid-state physics than in

10. Stephen Cole, "Scientific Reward Systems: A Comparative Analysis" (Paper read at Annual Meeting of American Sociological Association, Denver, 1971).

11. In some cases the number of citations received by scientists in various fields is partially influenced by the number of journals in each field in the *SCI*'s file. In 1971, sociologists have relatively few citations listed, in part because the *SCI* file contains only ten sociology journals—a small part of the literature in sociology.

elementary particles, we might expect that of two papers of equal impact in the two fields, the one in solid-state physics would receive a greater number of citations. While this position seems at first to be plausible, under closer scrutiny it does not appear logically sound. Although there are more citations being produced in solid-state physics than in elementary particles, there is also more work being done in solid-state physics. Consequently, there is a greater amount of total literature that can potentially be cited. Therefore, the probability of work receiving more citations simply because the specialty is larger does not logically hold. Furthermore, if size of specialty were related to number of citations, we would find a positive correlation between these two variables. Men working in larger specialties would receive more citations than men working in smaller specialties. We have evidence that, at least in physics, there is no relationship between the size of a specialty and the number of citations to the work of men in that specialty. According to the National Science Foundation, in 1966 there were 4,593 solid-state physicists and 1,833 elementary particle physicists.[12] However, our data on 1,308 university physicists show no significant differences in the rate of citation to physicists active in these two specialties. In the 1961, 1964, and 1965 editions of *SCI*, solid-state physicists had a grand mean of 17 citations while elementary particle physicists had a grand mean of 19 citations.

It may be true, however, that in a number of extreme cases there is a relationship between the size of a field and the number of citations to men working in that field. For example, in a specialty that has only a few scientists working in it, the number of citations to work produced would be limited by the small total number of citations to work in that specialty.

CONTEMPORANEITY OF SCIENCE

Papers in physics now have a half-life of no more than five years; that is, at least half the citations in the papers published in a given year are to work published in the five preceding years. We must take this into account in comparing the work of physicists who made their most important contributions at different times. Two papers which were originally of "equal" quality may have a different number of citations in the 1961 *SCI* if one paper was published in

12. See *American Science Manpower*, 1966, National Science Foundation, NSF 68–7, p. 183.

1941 and the other in 1959. This would not matter if the researcher was interested in the *current* significance of both papers. However, the researcher might want a measure of "quality" that is not time-bound. For this problem we developed a technique of weighting citations.[13] This consists basically of giving older citations greater weight than recent citations. Since 70 percent of the citations in a particular field (physics) are to work published within the preceding five years and 4 percent are to work published more than twenty years before, we gave a weight of 17 (70 percent divided by 4 percent) to work published twenty or more years before. Although weighting citations for their age seems to be substantively necessary, we found a high correlation ($r = .80$) between the total of weighted and unweighted citations. When we compare the number of weighted and unweighted citations to physicists' work in their three "best" years, we get an even higher correlation ($r = .96$). Since the number of a physicist's weighted citations is correlated so highly with the number of unweighted citations, substantive conclusions would probably not be affected if the weighting technique was not used.

If weighting is used, one more problem must be considered. Derek Price has suggested that "although half the literature cited will in general be less than a decade old, it is clear that, roughly speaking, any paper once it is published will have a constant chance of being used at all subsequent dates."[14] In the study of delayed recognition we found that in fact papers which were cited in 1961 have on the *average* roughly the same number of citations in 1966. Thus, at least for short periods of time, it is likely that Price's suggestion is correct. The weighting technique discussed above is not at odds with the model suggested by Price. The weighting technique is not meant to predict the *number* of citations received by papers in the past but to control for the increasing total number of citations. Due to the exponential growth in science, papers presenting important scientific contributions today are receiving many more citations than similar papers did in the past. A paper which receives five citations today is not among the most heavily cited. But a paper published in the nineteenth century which received five citations would probably have been among the most heavily cited papers. Therefore, in comparing work written in different periods we must standardize for the total number of citations being made.

13. Cole and Cole, "Scientific Output."
14. Derek Price, *Little Science, Big Science.*

INTEGRATION OF BASIC IDEAS

Widespread basic ideas are often utilized in papers without explicit citation to their well-known source. Who today cites that paper in the *Annalen der Physik* as the source of $E = MC^2$? For those scientists who have achieved one of the highest levels of recognition, eponymy, there may be a decline in the number of formal citations to their work. The "Mossbauer Effect" is an example of a recent contribution to science which has been thoroughly integrated into the common body of knowledge and is infrequently given formal citation. But let us examine such cases in terms of our measure. A scientist who has produced discoveries which lead to eponymous recognition probably will also produce other research of the first rank that will be heavily cited by the scientific community. Thus the research of most Mossbauers will be classified as "high quality," despite the fact that one of their outstanding achievements receives relatively few formal citations.[15] It is true, however, that integration of a discovery into the body of scientific knowledge may lead to errors in assessing the quality of that discovery through citations. The use of citations as a measure of quality does involve a degree of error. However, evidence presented above and below leads to the conclusion that such error, which may be substantial in individual cases, will not be significant in considering the work of any fair-sized sample of papers or authors.

STABILITY OF CITATION COUNTS OVER TIME

If citations to particular papers or scientists were to vary sharply from year to year, this might indicate that citation practices are somewhat random and that citation counts would not yield a reliable measure of quality. This is not the case. In a study of papers published in the *Physical Review* in 1963 we found a correlation of .72 between the number of citations received in 1964 and 1966, and even higher correlations for two adjacent years. Recently we have collected some data indicating that the total number of citations received by a scientist remains fairly stable over time. The data were on citations received by men and women scientists in 1961–70[16] (see table 1). All the correlations are relatively high. As we

15. Although the Mossbauer Effect is not as heavily cited as some other major discoveries, because of the degree to which it has been integrated into the general fund of knowledge, Mossbauer's lifework received a total of 53 citations in the 1965 *SCI*. Thus, using our measure, his work received roughly ten times the average and fifty times the modal number of citations in the *SCI*.

16. For a description of this sample see chapter 5 and Appendix A. Citation data do not exist for 1962 and 1963. We did not collect data for 1966 and 1968.

would expect in measuring most variables over time, the closer the points of measurement, the higher the correlation between the two measures. The correlation between the number of citations received by scientists in any two adjacent years is .90 or more. Even when we look at the number of citations received in 1961 and 1970, we get a fairly high correlation (r = .55). We may conclude that citations are a fairly stable measure of the impact of scientific work.

Table 1. Correlations for the number of citations received, 1961–70

	1961	1964	1965	1967	1969	1970
1961	—	.82	.77	.66	.58	.55
1964		—	.90	.82	.74	.72
1965			—	.87	.81	.79
1967				—	.93	.90
1969					—	.95
1970						—

N = 754

The problems with the use of citation counts as a measure of the quality of scientific output that have been discussed above all had a primary substantive basis. We now turn to a consideration of two problems which are primarily technical in nature, but nonetheless important to those who might make use of citation indexes.

CITATIONS TO COLLABORATIVE PAPERS

Citations to all single-author papers are recorded in the *SCI*. However, citations to collaborative work are listed only after the name of the first author in the collaborative set. Since many collaborative papers use an alphabetical arrangement of authors' last names, one might think that collaborators with names beginning with letters late in the alphabet would be misclassified if we counted only citations appearing after their names in the *SCI*. Data from our research suggest that, in fact, omission from the *SCI* of citations to this collaborative work does not present a formidable problem. For the sample of 120 physicists we have a full range of citation data. Not only did we have straight citation counts for each author, that is, the number of citations to work that the author produced alone and the collaborative work of his on which his name appeared first, but we also collected citations for all his collaborative work on which he was not the first author. These citations were obtained by looking up the author's collaborative papers in *Science Abstracts* and then looking up those papers on which he was not first author in the *SCI*. The correlation between a straight citation count and total citations (including citations to collaborative work on which the physicist was not first author) is .96. We also arrayed the 120 physicists in two ranked lists, one according to a straight citation count and the other according to total citations; the Spearman rank-order correlation coefficient is .85. Although the outcome was to some extent predetermined by the size of the zero-order correlations, we decided to make a final test of the relationship between straight counts and counts which included citations to collaborative work on which the physicist was not first author. We divided our sample of 120 physicists into two groups: the first group included physicists whose names began with letters in the first half of the alphabet; the second group, those with names starting with letters in the second half. For each group we calculated the percentage of total citations to either single-author papers or to first-author collaborative papers. By this technique we could estimate the extent to which scientists whose names begin with letters late in the alphabet were "deprived" of citations to work that they had actually helped produce. The finding indicates that there was little difference, on the average, between the two groups of scientists. Sixty-seven percent of the total citations to work produced by scientists in the first group and 71 percent of the total citations in the second group were to single-author or first-author collaborative papers. This small difference between the two groups suggests that the omission of collaborative citations to papers on which the author was not first among collaborators does not affect substantive conclusions. For the most part, differences that we did find were among scientists whose work was of the first rank. For example, Murray Gell-Mann had almost six hundred citations to his lifework in one volume of the index. When we looked for citations to collaborative research on which he was not first author, we found over one hundred additional citations to his work. While these additional citations add substantially to Gell-Mann's total of six hundred, they do not affect our classification of the quality of his work. However, when we want to study the quality of specific papers, we must look up collaborative papers under the name of the first author. Also, the researcher must be aware that because of the procedure adopted by the *SCI*, he may make errors in measuring the quality of work of a particular scientist.

CLERICAL PROBLEMS

Warren Hagstrom has recently pointed out other technical problems in the use of citation counts.[17] First, he notes that there are clerical errors in the list of citations. Second, since the names of cited authors are arranged by configurations of letters, it is possible that citations to two different scientists will be listed together. In other words, if there were two E. McMillans, one the Nobel physicist and the other a relatively unknown biologist, the citations to both men would appear together under the same name. Although these two problems make for greater inefficiency in collecting citation data, both can be handled. First, consider the problem of clerical errors in compiling the index. There is no reason to assume that these errors do not occur randomly throughout the index. Therefore, while the counts may be off slightly, there is no reason to believe that there are patterned errors in the listings. The second error is more vexing, but can be handled by careful compilation of the citation data. The index lists, along with the cited author, the names of his citers, the name, volume, and page of the journal of both the cited article and the journal in which the citation appears. Consequently, it is possible to identify the articles produced by the scientists that one is interested in. It is relatively easy to distinguish between scientists working in different fields; it takes more effort to distinguish between two physicists who happen to have exactly the same name.[18] Thus both problems do not materially detract from the value of the index as a measure of the quality of scientific output.

USE OF CITATIONS TO MEASURE OTHER CONCEPTS

Thus far we have discussed the use of citation counts to measure the quality of scientific papers. Citations can also be used to measure other important concepts in the sociology of science. Here we shall briefly mention three additional uses.

Recognition. One of the most significant ways in which scientists are rewarded is by having their work used by other scientists. Thus, the number of citations a scientist receives may be taken as an indicator of the amount of recognition his work has received.

17. Warren O. Hagstrom, "Inputs, Outputs, and Prestige of American University Science Departments," *Sociology of Education* 44 (Fall 1971): 375–97.

18. This is done by looking up the articles in the journals and using the listed institutional affiliation of the authors to separate the articles.

Diffusion. One problem that we address in this book is how location in the stratification system influences the diffusion of a scientist's discoveries. The number of citations received may be used as an indicator of the extent of diffusion of a scientist's work.

Utilization. Another problem that we shall analyze is how the location of a scientist in the stratification system determines who it is that utilizes his work. We use citations to measure utilization. However, here we are not so much interested in how many citations a scientist receives as we are in the characteristics of the citers. Nonetheless, the citation index provides us with a complete list of the citers of any particular scientist.

In using citations to measure several different concepts one must be careful to avoid tautologies. It would be easy to make a statement such as, "those scientists producing the highest quality work receive the most recognition," where all you have is a single citation count for each scientist. We have attempted to avoid such errors by careful use of indicators. Throughout this book we shall point out additional methodological problems involving the use of citations to measure key concepts.

CONCLUSION

The data available indicate that straight citation counts are highly correlated with virtually every refined measure of quality. Correlations between straight counts and weighted counts; between straight counts and those which take into account citations to collaborative research in which the author is not first author, are all greater than .80. Consequently, it is possible to use straight counts of citations with reasonable confidence. In some research situations it may be substantively more appropriate to use weighted counts or to take into account collaborative work, but the use of these refinements need rest only on the taste of the researcher, as opposed to methodological necessity.

It is clear that there are a number of definite problems involved in using citation counts as an indicator of the quality of scientific output. However, the value of this measure as a rough indicator of quality should not be overlooked. If we interpret small differences in the number of citations as meaningful, we would be in error as to how citations are useful measures. It would not be accurate, for example, to say that a scientist who receives six or seven citations to his lifework in the 1961 *SCI* has done better work than a scientist who receives five or six citations. In other words, citations should not be used as a fine measure of quality. However, there can be little doubt that large differences in the number of citations received by scientists do adequately reflect differences in the quality of the work.

3 PATTERNS OF STRATIFICATION IN AMERICAN SCIENCE

Social scientists of varying theoretical and political persuasions have concurred that there are no "classless," or unstratified, societies. Although it is perhaps possible to have such a society, an empirical example has not yet been found. Stratification appears to be one of those rare universal principles.[1] Whether we consider the sharply separated strata of a caste system or the more subtle forms of stratification found by cultural anthropologists in some primitive societies, we invariably find social inequality. Studies of social stratification have usually focused on inequalities derived from location in the occupational structure. While extensive research has been done on stratification within religious, political, and economic institutions and on inequality which is a consequence of sex, ethnic, and racial differences, comparatively little research has been done on stratification in other institutions. In this book we consider the stratification system of one institution that has not been widely analyzed—modern American science.[2]

Most scientists are aware that science is a highly stratified institution. In response to this stratification there have been, since the early 1950s and before, periodic complaints from various quarters of the scientific community about the skewed distribution of scientific resources.[3] Some "have-nots" have protested that power and resources are concentrated in the hands of a very small number of "haves." There are few disagreements that science is dominated by a relatively small elite; there are disagreements about the bases on which rewards in science are distributed. Given the sharp inequality

1. Kingsley Davis, *Human Society* (New York: Macmillan, 1949). Kingsley Davis and Wilbert E. Moore, "Some Principles of Stratification" *American Sociological Review* 10 (April, 1945):242–94.

2. For another discussion of stratification in science see Harriet A. Zuckerman, "Stratification in American Science," *Sociological Inquiry* 40 (Spring 1970):235–57.

3. Spencer Klaw, *The New Brahmins: Scientific Life in America* (New York: William Morrow, 1968); Daniel S. Greenberg, *The Politics of Pure Science* (New York: The New American Library, 1967).

in science of recognition and power, it is interesting that there has not been more conflict between "haves" and "have-nots." For certainly the level of conflict in science does not approach that which society has witnessed in other institutions. To understand why there has not been more pronounced social conflict in science we must examine the structure of the scientific stratification system.

This chapter is divided into two major sections. The first locates scientific occupations in the general occupational structure and maps out the various forms of recognition available to working scientists. The second deals with problematics in explaining patterns of stratification that actually obtain in science. It considers alternative modes of explaining inequality in science; discusses the role of universalism in the process of status attainment; examines the social processes which determine stratification; looks at the critical role that consensual agreement and intellectual and social authority play in maintaining an organized scientific social structure; and finally, concludes with a discussion of the level of social conflict in the social system of science. With this brief outline in mind, we move to a description of the stratification system of science.

PRESTIGE OF SCIENTIFIC OCCUPATIONS

Before describing the shape of stratification in science, we will consider the position of scientists in the American occupational structure. Some of the most powerful empirical generalizations in sociology have emerged out of the extensive literature on prestige hierarchies among occupations.[4] All studies of occupational prestige have found scientific occupations to be among the most prestigious. There were eight "scientific occupations" included in the 1947 and 1963 NORC studies of occupational prestige.[5] In 1963, scores of all occupations ranged from 94 for Supreme Court jus-

4. Robert W. Hodge, Paul M. Siegel, and Peter H. Rossi, "Occupational Prestige in the United States, 1925–63," *American Journal of Sociology* 70 (1964):286–302; Robert W. Hodge, Donald J. Treiman, and Peter Rossi, "A Comparative Study of Occupational Prestige," in R. Bendix and S. M. Lipset, eds., *Class, Status and Power: Social Stratification in Comparative Perspective* (New York: Free Press of Glencoe, 1966); Donald J. Treiman, "Industrialization and Social Stratification" in Edward O. Laumann, ed., *Social Stratification: Research and Theory for the 1970's* (Indianapolis: Bobbs Merrill Company, 1970), pp. 207–34; Alex Inkeles and Peter H. Rossi, "National Comparisons of Occupational Prestige," *American Journal of Sociology* 61 (January 1956):329–39; Albert Reiss, Jr., et al, *Occupations and Social Status* (New York: The Free Press of Glencoe, 1961).

5. A total of ninety occupations were studied. "Scientific occupations" were defined by the authors of the NORC studies.

tices to 34 for a shoe shiner. Scores for scientific occupations ranged from 92 for "nuclear physicist" to 78 for "economist." Clearly, the role of scientist is among the most highly valued in the entire spectrum of occupations. "Nuclear physicist" and "scientist" rank only behind Supreme Court justices and physicians in the overall scale. Although chemists and biologists rank slightly lower, their roles too can be considered among the elite occupations. There has been a slight increase in the evaluation of scientific occupations over time, but not a shift which is considered meaningful. In fact, the elite position of scientific roles is found in all societies for which empirical studies exist.[6] In the last twenty-five years there has been a growing awareness among the public of new scientific roles. The NORC trend studies show that in 1946 some 55 percent of the raters claimed that they "did not know" what a nuclear physicist did. By 1963 this proportion fell to 25 percent, although there was no indication that the public had more accurate knowledge of what a nuclear physicist actually did.[7]

We may conclude that within the overall occupational structure, scientists hold elite positions and there is relatively little differentiation among them when they are compared to the myriad of other occupations in the prestige hierarchy. In this work we are abstracting from the overall structure the various roles and statuses that scientists occupy and are viewing them under a microscope. In this way we can identify and analyze status differences which have little significance for the general population, but which have profound effects on men and women of science. For scientists, it is not only important to occupy a prestigious position in the national occupational structure, but it is perhaps even more important to occupy highly prestigious positions compared to other scientists within their own discipline. To most scientists, the relevant point of comparison is probably not their prestige compared to a "mail carrier," but their position compared to that of their fellow scientists.[8] There is a generic point here. Within all occupations there are considerable internal variations in prestige; country doctors do not have as great prestige among physicians as do the heads of large city hospitals.

6. Donald J. Treiman, "Occupational Prestige and Social Structure: A Comparative Analysis," unpublished manuscript. In fact, in the standard scale of occupational prestige developed by Treiman, "scientist" was the most highly rated of all occupational titles representing a large number of occupants.

7. Hodge, Siegel, Rossi, "Occupational Prestige."

8. For a discussion of the sociology of reference groups, see Merton, *Social Theory and Social Structure*.

Desire to achieve eminence within an occupation is a primary factor in motivating continued effort among workers once they have entered an occupation.

The identification of significant reference groups for scientists is an unexplored but potentially useful project. We would conjecture that occupants of low-status positions within the scientific community will be more likely than their eminent colleagues to use other, less prestigious occupations as points of reference. Many rank-and-file scientists may gain considerable satisfaction by comparing their positions to less prestigious occupations in the total American occupational structure. Among elites within science, the relevant comparative reference groups are probably their "peers" within their own and related scientific disciplines. Indeed, for those few scientists who reach the pinnacle of success within their own lifetime, the relevant comparative reference group may become the great figures in the history of science.

THE CONTOURS OF STRATIFICATION IN SCIENCE

Most often we think of the stratification system of science in terms of a hierarchy of individuals. At the apex of this hierarchy we find the handful of men who are virtually assured important positions in the history of science. Within the recent past we think of originators of new paradigms like Einstein and Planck, or great geniuses like Bohr, Fermi, and Pauling. These are the men who shape science and give direction to the work of their colleagues. Right below this handful of geniuses are the brilliant scientists who have been awarded the most coveted of scientific honors, such as the Nobel Prize or election to the various national academies of science. These men and the relatively few others who have received widespread renown constitute the elite stratum of scientists. They total probably no more than a few thousand out of more than one million active scientists in the world. Of course, with these positions of eminence come power, authority, and control of facilities and resources. These are visible men: the tip of the iceberg. But even many Nobel laureates and national academy members who are certainly members of the highest stratum of contemporary science may, when viewed in historical perspective, be seen as the bricklayers rather than the architects of science. The most eminent scientists reach their lofty positions through their contribution to scientific knowledge, the most highly valued form of activity in science.[9]

9. This section on the dual hierarchies in science has gained much from informal discussions with Professor Bernard Barber.

Another group that forms part of the scientific elite is scientific administrators. Though not having the prestige of men who entered the elite by virtue of their brilliant discoveries, administrators occupy influential positions in science. Acumen in administering science has frequently led capable but not extraordinary researchers to the highest levels of science policy-making. Scientists such as Ernest Orlando Lawrence, for one, achieved their reputations as much for their coordinating abilities as for their research skills.[10] From presidential science advisors to leaders of large research laboratories like Brookhaven, we find powerful scientists who have reached the highest strata of the institution of science through their organizational skills. These are the entrepreneurial scientists; the men who can get large-scale projects off the ground and see them through to completion. Administrative elites also serve as "gatekeepers" to many of the government-controlled resources.[11] They determine the criteria for awarding research grants, by selecting those who meet the criteria, and they are, consequently, significant forces in determining the foci of scientific attention. Much of the politics of science is handled by key administrators.

We are examining two social hierarchies in which different types of functional roles lead to positions of high prestige. On the one hand we have men who achieve eminence as a result of their contributions to idea systems; on the other hand we have men who, by virtue of occupying positions of responsibility, make contributions to the organizational structure of science.[12] In the history of science, the latter type of contributors are quickly forgotten, while the former live on for generations or centuries. This selective survival may well be a consequence of the way the history of science is written. It may turn out, however, that the key "administrators" of science are as important to the advance of idea systems as are the men who make the well-known breakthroughs. Science tends, of course, to hand out significantly more recognition for contributions to the idea

10. Nuel Pharr Davis, *Lawrence and Oppenheimer* (New York: Simon and Schuster, 1968).

11. For a discussion of the role of "gatekeepers" in science, see Diana Crane, "The Gatekeepers of Science: Some Factors Affecting the Selection of Articles for Scientific Journals," *The American Sociologist* 2 (November 1967):195–201; Harriet A. Zuckerman and Robert K. Merton, "Patterns of Evaluation in Science: Institutionalization, Structure and Functions of the Referee System," Minerva (January 1971):66–100.

12. Robert K. Merton and Harriet A. Zuckerman have recently been formulating ideas on the process of "institutionalization" of scientific disciplines and have begun to emphasize the important functions performed by the contributors to development of organizational structures. For a first piece of this work, see Zuckerman and Merton, "Patterns of Evaluation."

system of science than for contributions to its organizational development. Eponymous recognition is a case in point. Has the "father" of any scientific specialty ever been a scientific "entrepreneur"? The cases that we do identify will undoubtedly reflect the overlap between contributions to the idea and organizational systems of science.

The solidarity among the scientific elite made up of brilliant scientists and administrators has been described well by Harriet Zuckerman:

> These groups taken together can be construed as the national elite in science. They cover one or at the most two scientific generations, for, on the average, they were elected to the Academy [National Academy of Science] at 49 and are *en masse* 62 or so. Within each discipline, most have come up through the ranks together and have shared a set of common experiences. Twenty or thirty years ago, when they were just beginning their careers, disciplines were very small, making it possible for each to have known the others since his twenties. The present group not only shared post-doctoral years at the great European and American laboratories, at Bohr's Institute at Copenhagen, the Cavendish, or G. N. Lewis' at Berkeley, for example. They are also tied together by their experiences in World War II which were most significant for the physicists and slightly less so for chemists and biologists. These long associations make for a degree of social solidarity among members of this group which provides for integration of the sciences—even though disciplinary loyalties are strong. Not only do members of the elite share a set of commitments and interests, they also share the past.[13]

Beneath this elite are located a number of strata of lesser-known, less influential scientists. Perhaps at the top of the nonelites are those scientists who hold academic positions at high-ranked departments within their field. In American physics there are roughly 2,500 members of departments in which Ph.D.s are granted. Of these 2,500, perhaps, at the outer limit, 250 belong to the elite; the other 2,250 make up the strata right below it. Of course, there are distinctions made among these 2,250 based upon the prestige of the department to which they belong and the significance of their discoveries.

Thus far, we have considered only the small group of productive

13. Zuckerman, "Social Stratification," p. 239.

researchers, most of whom hold positions in academic settings. The mass of scientists, even those who spend some time on research, are remarkably unproductive. They produce few new ideas and rarely put anything in print. Although these scientists remain relatively invisible to the larger scientific community, they may well have achieved local prestige as excellent teachers of science or as the ranking members of their college or university science department.

There are parallel nonacademic hierarchies in science. Among government and industrial organizations, there is a loosely defined stratification along prestige lines. Scientists working for companies which give a significant proportion of their resources to basic research have greater prestige than those working for companies predominantly concerned with commercial technical applications. Companies such as Bell Telephone provide facilities and financial rewards as well as a significant quantum of recognition for research done in their laboratories. So among scientists working in industry there is a vertical hierarchy based on the "quality" of the organization that one works for. There is also, of course, a stratification system within each of these organizations.[14] Similarly, government agencies also have differential prestige. Working for the National Institutes of Health or the Atomic Energy Commission confers greater prestige and rewards than working for less distinguished organizations such as the Naval Radiological Defense Laboratory or White Sands Proving Grounds.[15]

Continuing this broad mapping of the contours of stratification in science, we can identify in an intuitive way distinctions in the prestige of scientific disciplines and scientific specialties. Although little empirical evidence exists, there seem to be social rankings of both disciplines and specialties. The social cues that we have to indicate such differences in prestige come partly from the flow of scientific manpower and partly from the subjective evaluation of ranking scientists. Although there is little hard evidence, it appears that, until the last few years, physics seems to have been the most prestigious of the scientific disciplines, followed roughly by chemistry, biology, astronomy, and geology. Physics seems to have had little difficulty in attracting many of the nation's top young scientific minds into its domain. Today conditions have changed. Physicists

14. Kornhauser, *Scientists in Industry;* Marcson, *The Scientist in Industry*.

15. Glaser discusses in detail the stratification of positions in a large government research organization, describing the various channels and criteria for social mobility. Barney Glaser, *Organizational Scientists: Their Professional Careers* (Indianapolis: Bobbs-Merrill, 1964).

have recently been complaining about the "flight from physics" and the difficulty of recruiting talented youngsters. Biology appears, on the other hand, to have gained enormously in prestige among scientists and aspiring scientists since the breakthrough by Watson and Crick in the mid-fifties. The relative distribution of national resources for the scientific disciplines also seems to follow definite trends.

Within a single discipline there also appear to be variations in the relative prestige of scientific specialties. Theoretical physics tends to carry more prestige than experimental physics; high-energy particle physics seems to be more prestigious than solid-state physics. Definitions of what constitutes a prestigious specialty appear to be a product of the assessment by the field as to the areas in which significant problems exist as well as the extent to which solutions of the problem are difficult, that is, require relatively scarce talent to solve.[16]

Finally, even nations are stratified in terms of scientific productivity. Price has gone so far as to say, only half-facetiously, that nations must publish or perish.[17] While smaller nations are beginning to produce a relatively larger proportion of the total number of the world's scientific discoveries, the international scientific community (in all fields) is clearly divided into the "haves" and the "have-nots." Among nations, the number of significant producers of discoveries can probably be counted on two hands. If we examine the distribution of Nobel Prizes by nation, as Zuckerman has done, we see that over the past twenty years four nations, the United States, England, France, and Germany have won 74 percent of all Nobel Prizes in science.[18] Evidently, whether we break down science into strata of individuals, consider it in terms of disciplines, or production of scientific ideas by nations, we see sharp inequalities in the distribution of prestige and rewards.

16. There is an on-going debate among sociologists as to what, in fact, are the determinants of the prestige of social roles. For a discussion of this debate see pp. 60–65 and 247–53.

17. Derek J. de Solla Price, "Nations Can Publish or Perish," *International Science and Technology* 70 (October 1967):84–90.

18. Zuckerman, *Scientific Elites*. Zuckerman has provided us with the following data. Since 1901, 268 laureates from 25 countries have been named. The United States has produced 81, Great Britain 48, Germany 45, and France 20, for a total of 194. No other country has had as many as 10 laureates (the Soviet Union has produced nine). After World War II, the United States has dominated the prize, receiving 67 of the 81 awards after 1943. By contrast, only 10 of Germany's 45 laureates were named after World War II. The nationality criterion used by Zuckerman is country of citizenship at the time of the prize.

Throughout this description of the stratification system of science we have made frequent use of the term "prestige." This may seem indicative of the sociologist's "abnormal" concern with such unimportant matters. We must make it quite explicit that it is the value system of science which determines the prestige accruing to various scientific positions. Of all scientific activities, original contribution to knowledge is the most highly valued. With few exceptions, the prestige attached to a man or a position is a function of the extent to which they have or are expected to make original contributions. Contribution to scientific knowledge is the underpinning of the stratification system. Concern with "prestige" is not empirically separable from concern with the people contributing to scientific advance.

Thus far, we have been describing the varying strata in the stratification system of science in terms of the prestige accorded the occupants. In American society, in general, rank is based upon a variety of factors, including the possession of money, property, and prestige, derived primarily from occupation, education, or inheritance. In science, unlike society at large, location in the stratification system is not determined by the amount of money earned. Although eminent scientists generally earn more than noneminent ones, income inequality in science is probably considerably less than in business. Furthermore, the income inequality that does exist is less significant than in other institutions because in science money does not have important symbolic value. In fact, the value system of science proscribes work that is motivated by the desire to earn money.

Just as income is not a significant differentiator of scientists, neither is property. Scientific equipment and facilities are hardly ever owned by individuals. The products of science, discoveries, are also usually not owned by individuals. Very few scientists patent their discoveries. The value system of science stresses the communal nature of scientific knowledge. Although there is no property in science as we commonly use the term, scientists do possess a surrogate form of property—recognition. Recognition in science is the functional equivalent to property; and the right to "recognition" is indeed an inalienable one for scientists.[19] Recogni-

19. Here we are quick to note, of course, that not all forms of property in general society are "inalienable." But, in science, recognition is truly such a possession, since without it as a rare and sought-after resource, the social structure of science would probably collapse. In the chapters that follow we stress the importance of this point.

tion by peers is probably the chief motivating force in modern science.[20] Because recognition is so important to scientists, there must be a reward system that identifies and honors scientific excellence wherever it is found. If the scientist desires to acquire "property," he can do so only through recognition by the system, since there are no other legitimate ways to obtain property in science. Furthermore, illegitimate avenues to recognition are scarce.[21] Obviously, some scientists have a lot of property, or have been richly rewarded by the social system, while others remain propertyless. The scientist who is located at Harvard or Brookhaven, who is the head of the National Science Foundation, who has won a Nobel Prize, or whose lifework has made a wide impact on his field, is a "rich" man in science. He has property. Correlatively, the scientist who has not been widely honored, who is located at a small, relatively unknown college, has little property.

FORMS OF SCIENTIFIC RECOGNITION

Recognition in science is granted in several important forms. As Merton pointed out some time ago, the institution of science has developed a reward system designed to give "recognition and esteem to those [scientists] who have best fulfilled their roles, to those who have made genuinely original contributions to the common stock of knowledge." [22] The graded forms of such recognition in science are many; among them, we want to examine three kinds.

Honorific awards

The first form of recognition is the granting of honorific awards and memberships in honorific societies. We have investigated the distribution of awards to members of the physical science community. Members of the twenty top-ranked physics departments listed more than 150 different awards after their names in *American Men of*

20. A growing amount of qualitative information tends to confirm this point. In interviews conducted with American Nobelists, Harriet A. Zuckerman has recurring testimony that peer recognition is a major motivating force behind the work of scientists of the first rank (Zuckerman, *Scientific Elites*).

21. Clearly, science is not analogous here to the economy. The vast literature on deviant behavior points to alternative means to "success" within the general economic social structure. See, among others, Merton, "Social Structure and Anomie," in *Social Theory and Social Structure;* Richard Cloward and Lloyd Ohlin, *Delinquency and Opportunity* (Glencoe, Illinois: The Free Press, 1960).

22. Merton, "Priorities in Scientific Discovery," especially pp. 639–47. The significance of recognition in the social institution of science has been further examined by F. Reif, "The Competitive World of the Pure Scientist," *Science* 134 (December 1961): 1957–62; Glaser, *Organizational Scientists;* Hagstrom, *The Scientific Community;* Crane, "Scientists at Universities."

Science (1960 edition). We collected data on 98 of these awards (see Appendix B).[23] Questionnaire data enable us to rank each award according to its prestige and visibility. Prestige scores were computed by taking the mean of ranks assigned by the sample of physicists; the visibility of each award was measured by the percentage of physicists who knew enough about it to rank it at all.

Most awards are not highly visible to the national community of scientists. Only 22 of the 98 awards studied were known by as many as half, and only 42 by as many as one-fifth, of the physicists returning the questionnaires. Evidently, a large number of the awards which recipients take enough pride in to list after their names in AMS are local honors which, though not a part of the national reward system, may nevertheless confer local prestige. For this investigation, we adopted the convention that awards which were unknown to more than 80 percent of the physicists would be regarded as thoroughly parochial and excluded from further analysis.

Several aspects of the distribution of honorific awards are noteworthy. Of the 42 awards meeting the criterion of national visibility—familiar to at least 20 percent of the physicists—two stand out above all the rest. They are, of course, the Nobel Prize (with a prestige score of 4.98 of a possible 5.0) and membership in the National Academy of Sciences (4.22). What is more, physicists holding these most prestigious honors monopolize all the other high-prestige awards (awards with scores of 4.00 or better). In the sample of 120 American university physicists, there was not a

23. The list of 98 awards was assembled from *American Men of Science* and *Physics Today*. Many of the awards listed in *AMS* were of such limited local significance that they could not be identified and were omitted from the study. All the awards listed in *AMS* which could be identified were included. Added to this list were awards which appeared frequently in *Physics Today*. Thus, our list of 98 represents a large sample of all awards granted to physicists.

Since it was not feasible to ask each physicist to rank 98 awards, we used five different forms of the questionnaire. Ten awards were included on all five forms. Since the difference between the scores of these awards which appeared on all five forms was statistically insignificant, we conclude that the score received by each award is representative of its prestige among academic physicists. As an example of the closeness of the ratings on the five forms, membership in the National Academy of Sciences received ratings of 4.28, 4.32, 4.03, 4.24, and 4.27 on the five forms. The wording of the question was: "The following thirty awards represent a sample of several kinds of awards. For those which are known to you, we ask that you indicate your judgment of its prestige by circling one of the five rankings. You may not have heard of many of these awards, as most are not widely known. If you have heard of an award, but do not know enough about it to evaluate its prestige, please circle No. 6. If you have never heard of the award, circle No. 7. The circling of either 6 or 7 provides useful information as a ranking since it will indicate which awards are least known among physicists."

single recipient of the Fermi Award, Royal Astronomical Society Gold Medal, Albert Einstein Medal, Fritz London Award; and not a single member of the Académie Française or Royal Society who had not received a Nobel prize or was not a member of the National Academy of Sciences. However we interpret it, the fact is that the awards conferring topmost eminence in physics are closely confined to a small group of physicists.

As table 1 shows, the prestige of the highest award received by physicists is correlated with the total number of their awards ($r = .70$).

Table 1. Prestige of highest award by total number of awards received

	Number of Awards					
Prestige of highest award	4 or more	3	2	1	Total	N
Nobel laureates	45%	36%	19%	–	100%	(11)
Member of National Academy of Sciences	27	23	36	14	100	(22)
Awards with scores of 3.00–3.99	–	12	55	33	100	(33)
Awards with scores of less than 3.00	–	–	17	83	100	(24)

NOTE: Of this sample of 120 university physicists, thirty had received no awards. For a description of the sample, see Appendix A.

By way of anticipation, we note here that the total number of honorific awards correlates as highly as the prestige of highest award with almost every other indicator of recognition of scientific accomplishment. In due course, we shall examine the distribution of total number of honorific awards among various types of research physicists.

Although awards and prizes are symbolically an important part of the reward system of science, they are not widely distributed. In our national sample of physicists teaching in departments granting Ph.D.'s, itself an elite group of roughly 2,500 physicists, 72 percent had received no awards whatsoever, not even a graduate fellowship. Another 15 percent had received one award, in most cases postdoctoral fellowships. Even among the 632 physicists employed by the top twenty departments of physics in 1960 (at the rank of assistant professor or higher), only one-third listed any awards in

American Men of Science.[24] That only a few scientists have received honorific awards is not surprising. If it were otherwise, the awards would lose much of their social meaning.

The functions of awards

In the past, when sociologists of science have discussed the scientific reward system, they have concentrated upon the most renowned and most widely known forms of personal recognition. For example, Merton often mentions eponymous fame and prestigious awards in describing the reward system of science;[25] and Zuckerman has done an extensive analysis of Nobel Prize winners.[26] Since recipients of Nobel Prizes and members of national academies are the men who count most for scientific advance, it is not unnatural to concentrate attention on the reward system for this superelite. Here, however, we turn from an analysis of prizes like the Nobel Prize, which are, after all, among the most notable, dramatic, and glamorous exceptions in science, to inquire into the entire spectrum of scientific awards. We are primarily interested in the functions of awards of varying prestige and visibility.

Ideally, awards should act to increase the visibility of successful scientists who serve as role models for other scientists. Awards should spotlight work by scientists who have lived up to the norms of science and have, in the process, made original contributions to their fields. By focusing attention on excellence and honoring it, awards should reinforce the patterns of behavior which led to the production of superior research. From another point of view, awards should increase the prestige and visibility of the granting organizations by linking them with the great men of contemporary science. Zuckerman has noted, for example, how organizations that are in any way connected with Nobel laureates, claim them as their own in order, we assume, to enhance their own prestige.[27] Finally, awards should motivate the working scientist to focus on important problems—the solutions to which are likely to be rewarded.[28] Awards in general, then, should serve significant reward functions

24. The year 1960 is used because a later edition of *AMS* was not available when the data on awards were collected.

25. Merton, "Priorities in Scientific Discovery."

26. Zuckerman, *Scientific Elites*.

27. Harriet A. Zuckerman, "The Sociology of the Nobel Prizes," *Scientific American*, 217 (November 1967):25–33.

28. See, among others, Merton, "The Matthew Effect"; James D. Watson, *The Double Helix* (New York: Atheneum, 1968).

for the individual scientist, the granting organizations, and the entire scientific community.

As we pointed out above, it is not surprising that only a few scientists have received honorific awards. What is more surprising is the fact that most scientists do not know of many awards. If few scientists have heard of many awards, how can the latter fulfill the functions which we have just discussed? What functions do awards in fact serve for the recipients, the organizations granting them, and the scientific audience?

Awards as motivators. Awards with different prestige serve as motivators for scientists who are located at different levels in the stratification system of science. The Nobel Prize probably has no motivational significance for most working scientists. In any realistic sense of motivation, it is irrelevant because the average scientist knows he has no chance of gaining the honor. For the few, such as biologist James D. Watson, the desire to win a Nobel Prize may serve as an important motivating factor in their research.[29] But even among these would-be laureates, there exist so many other sources of motivation for continuing to produce high-quality re-

29. See Watson, *The Double Helix.* Competition among scientists for honorific awards remains a relatively unexplored yet potentially interesting area of research. From the Watson account it appears evident that for men of the first rank, working at the research front, competitive aspects of problem-solving are of importance for the rapid advance of science. For Watson, once the strategic problem of finding the structure of DNA was located, beating Linus Pauling to the solution—or as Watson puts it—handing Linus his first scientific "defeat," was central to his entire research effort. Competition for scientists has a close analytic relationship to the problem of priority in science. It is a well-known fact that many scientists are concerned about being anticipated. Most often this implies that they fear that other scientists, working on the same scientific problem, but who remain unknown, will reach a solution to the problem first. However, in contemporary science, one's competitors are not necessarily "invisible." A scientist may be well aware of the other scientists who are working on the same problem. Moreover, they may know the state of their "competitors' " research. Priority disputes are likely to have different consequences for both the individual and the social system when competition is open, in the sense that one knows one's competitors, in contrast to the situation where the scientist's competition remain "invisible." When scientists are aware of their competitors it may turn out that "game" elements enter into the research effort. Consequently, reactions to having been "beaten" to a solution may be different than reactions to anticipation where the scientist did not realize that he was involved in competition. Further, we might suggest that the seeking of competition is a stratified phenomenon. Scientists of the first rank may be more likely to focus their energy upon similar important problems, since the payoff for solution is clearly seen by them. These are not the scientists who are going to play it safe. It would be of value to inquire into types of scientists who seek out strong competition.

search that awards, in themselves, would seem to play only a minor role.

Other awards of less prestige may provide some motivation for scientists at lower levels in the stratification system. Young scientists, working in mathematical physics, might have it in the back of their minds that they might win the Dannie Heinneman prize, which is given only to young scholars. For graduate students, fellowships may act as motivators, since they have a realistic chance of gaining the honor. At each level in the social system, three conditions seem to be necessary for awards to have motivational significance. They must be known to the stratum; they must carry some degree of prestige; and they must be within reach.

Awards as validators of role-performance. Distributing honorific awards is one way in which the social system of science validates past performance and testifies to the potential of a scientist. Awards serve this function for both eminent and noneminent scientists. For young noneminent scientists, local rewards are evidence of future potential. When a graduate student, looking for his first academic position, places on his vita "invisible" graduate school awards, he is communicating the fact that his work as a student has been recognized as outstanding. Graduate school awards are age-graded indicators of potential and are soon removed from the vitae of scientists who go on to publish significant work and receive more prestigious and visible awards.

For eminent scientists, awards are indicators of past successes and increase the probability that the recipients will have access to resources and facilities. The scientist who holds a Lasker Award or a Rumford Premium is more likely to receive research grants and future awards than the scientist who has received no honors.[30] Thus awards channel resources to those scientists who have in the past been most successful.

Function of awards for granting agencies. Award-granting agencies confer prestige on themselves as well as on the scientists they honor, when the recipient of their award is either a well-known scientist or later becomes an eminent scientist.[31] Scientific societies, by honoring superior research, also facilitate the communication of

30. See Merton, "The Matthew Effect."
31. Zuckerman, "Nobel Prizes."

new ideas. Work that is rewarded becomes highly visible, if the granting organization is itself well known. And work that is widely known may be widely used.

Men and women who have been honored by prestigious scientific organizations become statesmen and diplomats of science. They symbolize scientific achievement. Thus award-granting societies, by honoring outstanding role-performance, also function to reinforce the normative structure of science. The men and women who have been honored have lived up to the most important norm of the scientific community; they have made substantial contributions to their fields. By rewarding high-quality work, the granting organization reinforces patterns of work that lead to significant scientific discoveries.

The visibility of scientific awards

We have noted only a few functions that awards fulfill in the social system of science. While an award need not have high visibility in order to serve a purpose, it remains problematic as to why some awards are widely known while others remain relatively obscure.[32] Table 2 presents a matrix of zero-order correlations for several characteristics of honorific awards. All of these variables are correlated with the visibility of awards. The variable most strongly correlated with visibility is the prestige of the award. Low prestige awards

Table 2. Intercorrelations of several characteristics of honorific awards

	V	P	Q	M	R	S
V = Visibility of award	–	.74	.35	.22	.36	.50
P = Prestige of award		–	.52	.18	.22	.64
Q = Quality of recipients' research			–	.02	–.04	.41
M = Monetary reward				–	–.01	–.12
R = Total number of recipients: 1955–1965					–	–.11
S = Scope of award						–

N = 65

NOTE: Fellowships and several awards on which a complete data set was unavailable were excluded from the analysis of the 98 awards we studied.

32. For an enumeration of the sample of awards, their visibility and prestige scores, see Appendix B.

rarely become highly visible. "Scope of award" is also strongly correlated with visibility. Scope was defined as the size of the pool from which recipients are drawn. Thus awards with the broadest scope are given for general excellence in any field of science. An award with a narrow scope would be one given for publication of an outstanding paper in a particular journal.[33] The number of recipients eligible for the first type of award is, of course, much greater than that eligible for the second. The broader the scope of the award, the greater its visibility. The quality of recipients' research as measured by citations, the amount of money attached to the award, and the number of recent recipients all are moderately correlated with an award's visibility.

We wanted to know the independent effect of all of these variables on visibility. To this end we have constructed a path model which we believe most closely approximates the way in which the variables of Table 2 influence the visibility of awards.[34] This model is presented in figure 1. The only variable having any substantial independent influence on visibility is prestige. All the other variables influence visibility indirectly through their influence on prestige.

The data of figure 1 present several interesting problems. Perhaps most important is the fact that the quality of recipients' work has no direct effect on visibility and only a moderate effect on prestige. The lack of a stronger relationship between the quality of recipients' work and visibility of awards at first appears anomalous since we know that the most visible awards in science have been given almost exclusively to scientists of the first rank. The research produced by recipients of top awards like the Nobel Prize is among the

33. There were five coding categories of "scope": (1) excellent performance in one of several scientific disciplines; (2) excellent performance in a single scientific discipline; (3) excellent performance in a specific specialty of one discipline; (4) excellent performance in a subspecialty of one discipline; (5) publication of outstanding paper in a local organization journal, or for outstanding teaching.

34. The path model presented takes a simple recursive form. We estimated models with reciprocal paths between prestige and visibility of awards. These models did not as well represent what we felt to be the actual process through which awards gain prestige and visibility. In this case the simple model was also the best. It is, of course, probable that over time prestige and visibility are mutually reinforcing. To estimate the influence of visibility on prestige of awards would require data over time. For a discussion of path analytic techniques, see the essays in Hubert M. Blalock, ed., *Causal Models in the Social Sciences* (Chicago: Aldine, Atherton, 1971); O. D. Duncan, "Path analysis: sociological examples," *American Journal of Sociology* 72 (July 1966): 1–16; K. C. Land, "Principles of Path Analysis," in Edgar Borgatta, ed., *Sociological Methodology* (San Francisco: Jossey-Bass, 1969), pp. 3–37.

most visible and widely used.[35] We would not expect that the quality of work produced by recipients of low-prestige awards would be equal to that produced by laureates and academy members. Yet, even if we do not control for prestige of awards or any other independent variable, the quality of work of recipients explains less than 10 percent of the variance on visibility of awards. Indeed, the path analysis suggests that quality of recipients' research has virtually no direct influence on visibility.

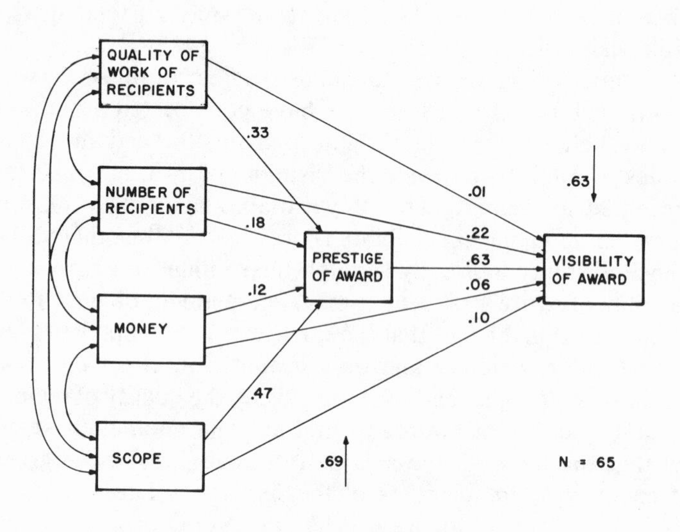

FIG. 1 VISIBILITY OF AWARDS ⟶

The key to understanding this result lies in the fact noted above: that formal recognition has a direct effect on only a limited number of scientists, since honorific awards are given only for high levels of scientific performance. These data do not imply, then, that highly visible awards are given to scientists who have not made outstanding contributions to scientific progress. Rather, awards which remain relatively unknown to the physics community are also being given to men who produce work of excellence. Only two examples are necessary to illustrate this point. The average number of cita-

35. For a discussion of the relationship between the visibility of individuals and the prestige of their highest award, see chapter 4.

tions to the work of recipients of the Nobel Prize (visibility score of 100) in the last ten years was 199 in the 1965 *Science Citation Index*[36] (see table 3). In contrast to the Nobel Prize, the National Medal of Science, with only moderate visibility—33 percent—had recipients with a mean number of citations of 154 for the same period of time. When we consider that the mean number of citations to the work of all scientists listed in the 1965 *SCI* is 6.08, it becomes clear that the quality of work produced by recipients of all awards is far above average, but does not vary directly with increasing visibility scores. Therefore, the visibility of awards cannot be explained

Table 3. Visibility and prestige of awards and average number of citations to recipients

Name of award	Visibility score	Prestige score	Mean no. of citations (1965 *SCI*)
Nobel Prize	100	4.98	199
National Academy of Sciences	95	4.22	127
Oliver E. Buckley Solid-State Physics Prize (American Assn. of Physics Teachers)	73	3.65	168
American Physical Soc. Prize	55	3.40	109
Atoms for Peace Prize	50	3.80	88
National Medal of Science	33	4.02	154
Fellow of the American Philosophical Society	20	2.79	115
Ballantine Medal (Franklin Institute)	14	2.60	60
Eugene Cook Bingham Medal (Society of Rheology)	10	2.80	75
Bruce Gold Medal (Astronomical Society of the Pacific)	6	3.20	95

by the varying quality of research of their recipients. Additional comparisons of citation rates to recipients of a selected sample of awards with different visibility and prestige scores appear in table 3. These data suggest that honorific awards (we do not here include graduate fellowships), regardless of their visibility are given almost exclusively to scientists who, by any standards, are of the first rank in their fields.

36. The data presented here include citations to scientists working in all fields of science, not just in physics. Since chemists often receive more citations to their life work than physicists, these figures seem at first to differ significantly from those presented earlier for physicists who had received the prize. Also the figures used above were based upon an analysis of the considerably smaller 1961 *SCI* file.

Figure 1 shows that the only variable other than prestige to have an independent effect on visibility is the sheer number of recipients. Apparently not only must awards be of high prestige to be known, but the community of physicists must be given a steady dose of recipients. Take, for example, a problem which confronts the National Academy and Columbia University in awarding the Barnard Medal. This medal, awarded once every five years to a single scientist, has been given to some of the most celebrated scientists in this century, among them Rayleigh, Roentgen, Rutherford, Einstein, Bohr, Frederick Joliot, Irene Curie, Fermi, and Rabi. Virtually all of them have been recipients of the Nobel Prize, either before or after the Barnard Medal. Yet the Barnard Medal is practically unknown to the community of physicists, having a visibility score of 6.[37] We would conclude that one reason for the low visibility of the Barnard Medal is its infrequent presentation.

It appears, then, that visibility is affected by the number of recipients of a prize. Worth noting is the slight curvilinear relationship that exists between number of recipients and prestige of awards. Those awards that are given to large groups of scientists, especially the postdoctoral fellowships, have lower prestige but as high visibility as those awards which are given to a moderate number of scientists. It would indeed be difficult to estimate the optimum level of "exposure" for an award. However, if a great number of scientists receive an honorific award, it may become insignificant that anyone has received it.

A number of scientists whom we interviewed said with little hesitation that awards with high visibility and prestige would turn out to be those which also had large sums of money attached to them. The data of figure 1 suggest that the scientists were mistaken. Monetary awards associated with prizes had very little independent effect on either visibility or prestige. An explanation of the belief in the power of money may be found in the fact that a number of the most prestigious and visible awards did, indeed, have large sums of money attached to them. Most notable among these are, of course, the Nobel Prize, with about $100,000 going to its recipients, and the Fermi Prize, which rewards its winners with about $75,000. Both

37. A visibility score of 6 may be viewed as functionally equivalent to zero since a number of fictitious awards included in the survey were "known" to as many as 6 percent of the sample. For a further discussion of the ratings received by fictitious awards, see chapter 6.

of these awards are enormously prestigious and highly visible. What many observers overlook is that there are other awards of varying degrees of visibility and prestige which also involve handsome sums of money. Take, for example, the Atoms for Peace Prize, good for approximately $40,000 to its winners, yet known to far fewer physicists (51 percent) than either the Nobel or Fermi prizes. Correlatively, there are also awards of considerable prestige and widespread visibility which do not reward recipients with any money. To mention only a few: membership in national academies, the Einstein and Max Planck medals, the Fritz London award, and the Dannie Heinemann prize. In short, the data presented here suggest that the visibility and prestige of an award is not really dependent upon the amount of money associated with it. Money alone will not make a prize either well known or prestigious.

Occupational position

Since prestigious awards and membership in honorific societies are in such short supply, they are inadequate in providing recognition of lesser degrees of scientific accomplishment. A second form of recognition for scientific work, one which is more widely distributed, is appointment to prestigious jobs. This form of recognition may be symbolically less significant than receipt of formal awards, but, from a practical point of view, it is the most important. Few scientists are fortunate enough to have their work recognized by receipt of awards, and, as we shall see, few scientists are rewarded by having their work widely used; but most scientists have jobs. Among academic scientists, recognition is granted through the appointment to professorships in prestigious departments. There is a great deal of consensus among scientists as to the prestige rankings of academic departments.[38] Practically all physicists agree that Columbia and Harvard have excellent physics departments, just as most sociologists would agree that Berkeley has one of the top departments of sociology. A large majority of the scientific elite are recognized by appointment to prestigious academic departments. Positions at some of the leading research laboratories or institutes, such as Brookhaven or the Lawrence Radiation Laboratory, have almost as much prestige as professorships at the top departments. Some leading scientists who could easily get jobs in top departments

38. Allan M. Cartter, *An Assessment of Quality in Graduate Education* (Washington, D.C.: American Council on Education, 1966).

prefer to work at research laboratories either because they find the atmosphere there most conducive to research or because the equipment they need for their work is more readily available.

Very few jobs in other organizations will bring a scientist as much prestige as professorships in top departments or positions in top labs. Almost all industrial and government jobs offer less prestige, although, as was pointed out, there are prestige gradations in both government and industrial scientific organizations. While most scientists have jobs, few confer national prestige. Of the roughly twenty-five thousand physicists belonging to the American Physical Society, probably no more than 10 percent have positions which confer any form of national recognition.

Visibility

The third form of recognition we consider here is the most widespread and, in the view of Alan Waterman, operates as a greater incentive for scientists than more formal recognition, like awards and prizes. This is "the kind and degree of attention [one's] research receives from the scientific community."[39] In this book, we introduce an additional and perhaps more exacting indicator of feedback from the scientific community. We have collected data on a sample of nearly 1,300 university physicists. At one point, the sample was asked to describe the extent of their first-hand familiarity with the work of each of a stratified random sample of 120 physicists; if they had not read the scientist's work, they were asked to indicate whether they had heard of him at all (see Appendix A). We have adopted the percentage of the community of physicists who reported knowing any part of a physicist's work at first-hand as a measure of the visibility of the scientist. To be widely known to one's colleagues, to have a widespread reputation, is a form of honor as well as one type of social reinforcement. In the next chapter, we shall examine the determinants of the visibility of these physicists in order to discover the process through which professional reputations are made.

In order for a scientist to receive this third type of recognition—citation of his work, or high visibility as a result of contributions to his field—he must usually first publish scientific work and then have such work favorably evaluated by colleagues. Since few scien-

39. Alan Waterman, *Science,* 151 (January 1966):61–64. For discussion of the same point, see John Ziman, *Public Knowledge* (Cambridge: Cambridge University Press, 1968).

tists publish many papers and even fewer have their work favorably evaluated, this form of recognition, as the two previous ones, is not widely dispersed. Price has estimated that for every 100 scientists who produce one scientific paper, there are only 25 who produce two, 11 who produce three, and so on.[40] This pattern, first identified by Lotka, closely approximates an inverse square function: the number of scientists producing n papers is roughly $1/n^2$. To put the matter simply, only a small proportion of scientists produce the bulk

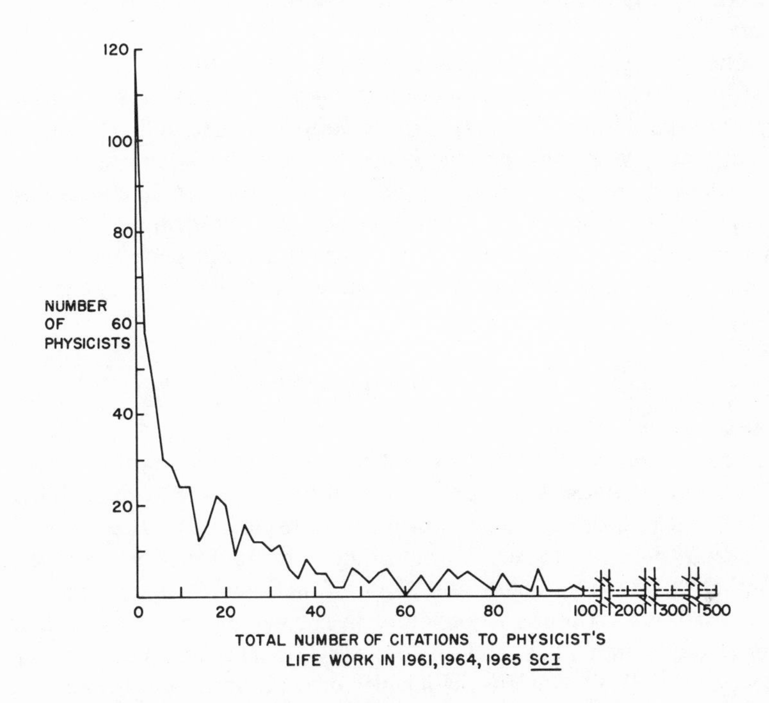

FIG. 2 DISTRIBUTION OF CITATIONS TO UNIVERSITY PHYSICISTS

of science which emerges from the scientific community. When, say, 10 percent of the individuals in an institution produce 50 percent of its total output, we may reasonably conclude that a sharp stratification of productivity is present.

If we consider the use of scientific research, we find a similar skewed distribution. One indicator of use is citations. Figure 2

40. Price, *Little Science, Big Science,* p. 43.

clearly indicates that most scientists in at least one discipline, physics, rarely have their work used. Some 275 out of 1,308 university physicists, or 21 percent, had received less than an average of one citation a year to their lifework. Many of these scientists are, of course, totally unproductive and thus have no potential for being cited. Almost half of the physicists have been cited, on the average, fewer than five times a year. Inspection of figure 2 suggests that only a small fraction of the total number of active scientists receive recognition which takes the form of widespread use of their work.

The different forms of recognition we have discussed may be divided into two types: recognition through position and recognition through reputation. The first form is quite concrete and consists of occupying prestigious positions in the scientific social structure. Recognition through reputation is more amorphous. It consists of the visibility of the scientist, and his colleagues' informal evaluation of his work. The two types of recognition are not perfectly correlated. In the next chapter we shall take a detailed look at the determinants of both.

FUNCTIONAL AND CONFLICT THEORIES OF STRATIFICATION

The functional theory

Before examining the processes through which recognition is distributed to scientists, it might be useful to discuss the applicability of general theories of social stratification to an analysis of science. The two theories of greatest importance are the functional and the conflict theory. The Davis and Moore functional theory posits that rewards are distributed to positions in society on the basis of their functional significance and the scarcity of available talent to fill the position.[41] Those jobs which perform the most vital functions for society and are most difficult to perform are the most heavily

41. Davis and Moore, "Principles of Stratification"; Davis, *Human Society*. For a critique of this theory see, among others, Melvin M. Tumin, "Some Principles of Stratification: A Critical Analysis," *American Sociological Review* 18 (August, 1953):387–94; Richard L. Simpson, "A Modification of the Functional Theory of Stratification," *Social Forces,* 35 (December 1956); Walter Buckley, "Social Stratification and the Functional Theory of Social Differentiation," *American Sociological Review* 23 (August, 1958):369–73; Dennis H. Wrong, "The Functionalist Theory of Stratification: Some Neglected Considerations" *American Sociological Review* 24 (December, 1959); Arthur L. Stinchcombe, "Some Empirical Consequences of the Davis-Moore Theory of Stratification," *American Sociological Review* 28 (October 1963):805–808.

rewarded. The purpose of unequal rewards is to motivate talented individuals to enter positions where they are most needed.

How might this theory be applied to science? Unlike society at large, science is difficult to describe as a series of discrete positions hierarchically arranged. We could differentiate academic scientists according to the type of institution in which they worked. University professors would be at the top of the hierarchy and teachers at junior colleges at the bottom. There is, however, considerable overlap in the functions performed by occupants of these positions. All those holding the positions are teachers of science. People at the bottom of the hierarchy are expected to do less research than people at the top. It would also be difficult to locate nonacademic positions in the hierarchy. Some industrial scientists, for example, have as much prestige as university professors.

Prestige differentials that do exist among positions probably are linked to the functions performed by the occupants. The more the occupants are expected to carry out pure research, the greater would be the prestige. The functional theory claims that those positions which serve the most important functions for society have the highest prestige. The theory has been criticized for its seeming inability to specify how functional importance can be determined. Functional significance, meaning necessity for the continuing operation of the society, is probably not a determinant of the prestige of occupations. Defining functional significance in this way, we would have to conclude that the function performed by sanitation men is probably at least as important as that performed by a college professor. Functional significance as a criterion for prestige makes sense only if it means the value that society places on particular tasks. Society places higher value on the task performed by a college professor than on that performed by a garbage collector.[42] In science the task which is most highly valued is research contributing to the advance of scientific knowledge. Thus, in science the more the occupant of a position is expected to contribute to "pure" research, the greater the rewards attached to the position.

The second and more important part of the functional theory of stratification is the hypothesis that the prestige of an occupation depends upon the scarcity of available talent capable of filling the position. Thus there are fewer people with the training and ability necessary to be a college professor than there are people capable of being a sanitation man. The functional theory applied to science

42. The crucial question, of course, is how people come to have such values.

would argue that there are few scientists with the skill needed to make contributions to scientific knowledge and that most scientists do not contribute to progress through their research.

Are differential rewards necessary to motivate talented scientists to seek out positions requiring research? Just as this question is difficult to answer for society at large, it is difficult to answer for scientists. We must keep in mind that the rewards scientists receive are more often recognition than material rewards. If material rewards in science were more evenly distributed, there probably would not be a sharp decline in motivation. However, failure of the system to provide recognition for work well done would probably have more serious negative consequences. Some scientists, no doubt, would continue to work hard on their research even if the norms prescribed that the researcher must remain anonymous. These scientists have the "sacred spark." They are motivated by an inner drive to do science and by a sheer love of the work. Others could probably not generate the self-discipline required to do science without the expectation of personal recognition. Most scientists are motivated, of course, by some combination of love for the work and desire for personal recognition.

Failure to recognize individual contributions would probably have negative consequences for science even if it did not lead to reduction in individual motivation. The system benefits in several ways from having "stars"—scientists who have received much recognition. Stars serve as role models for younger scientists. If all work was anonymous, students would not have men to emulate who embody the values of science. More important, students would not know whom to study with. It is important that the brightest students get to work with the men who are on the research frontier.

If there were no differentials in distribution of recognition, research resources would be more evenly distributed. Some scientists would surely see this as a positive good. But from the point of view of the system, it is probably advantageous to have the bulk of research resources go to the men who have the greatest scientific ability. The reward system in science serves to single out those who have been successful in the past and to give them the resources to make additional discoveries.

The hypothesis that stars serve important functions for scientific social systems would seem to be supported by the current state of sociology. Most of the current stars in sociology are men over the age of sixty; few are under fifty. This is probably a result of the current intellectual state of the discipline. Sociology is in a highly inductive phase in which contributions are almost by necessity minor rather than major. Yet despite the fact that few stars under the age of fifty have emerged, the field seems to be trying to create new stars. This results in the phenomenon of young scholars having wide reputations prior to their publication of work of major significance. The field seems to be searching around for young people who can take the mantle from the retiring stars. This analysis, of course, is based upon our impressionistic view of sociology. If empirical data could be collected to support the analysis, it would offer striking support for the hypothesis concerning the functional necessity of stars.

The argument that stars are needed to fulfill some of the functional requirements of scientific social systems poses some important questions for the sociology of science. If, in fact, scientific fields without naturally emerging stars are forced to create "synthetic" stars, this would be a crucial example of how the organization of scientific work is influenced by sociological factors. It is possible that scientific disciplines need stars for sociological as well as intellectual reasons. Later in this chapter we shall discuss some additional sociological needs fulfilled by stars.

The conflict theory

The principal intellectual antecedent to conflict theory lies in the work of Marx and Engels. Recently, however, sociological theorists of this persuasion have freed the orientation from its long-standing dependence upon economic relationships as the determinant of all other relationships. The theory, to be all too brief, takes as its starting point the basic assumption that social change and social conflict are present and are in continual process in all sectors of society. In analyzing stratification systems, the underlying assumption in the orientation is that social institutions invariably are structured by conflicting sets of interests. Through control over resources, facilities, and rewards, some members of the society are in positions of superordination, while others are in positions of subordination. In short, all institutional structures may be characterized by dominant and coercing elements juxtaposed against submissive or coerced elements; social structures may be described in terms of the conflict between the oppressed and the oppressors. Dahrendorf succinctly summarizes this position: "In institutional terms, this means that in every social organization some positions are entrusted with a right to exercise control over other positions in order to ensure effective coercion; it means, in other words, that there is differential distribution of power and authority."[43] The distinctive feature of the orientation relative to functional analysis is that conflict theory focuses on the social relationships which emerge over conflicting interests *independently* of the subjective perceptions of the participants. To conflict theorists, the elite in any institution must have interests different from and opposed to those of nonelites. Functional analysis does not assume any inherent conflict of interests among people occupying different positions in the stratification system.

A basic difference between conflict and functional theorists is their assessment as to why elites monopolize rewards and prestige. Where functionalists see this outcome as a result of the significance of the tasks performed by the elite, and their unique abilities in performing such tasks, conflict theorists see the uneven distribution of rewards as a result of the unequal distribution of power. Elites use the power at their disposal to extract an unfair share of available rewards at the expense of the nonelites. They also utilize their control of the media to perpetuate a value system which will preserve their position of superiority and lend legitimacy to their "rule." The ruling elite fosters in the subordinates the ideology that their station in life is a result of personal failings rather than structural conditions.

To what extent is the conflict theory of stratification applicable to science? In science, as in society at large, there are definite differentials in the distribution of power. Directors of laboratories, advisors to government agencies, editors of journals, and chairmen of graduate departments have considerably more power than the average scientist. But to assert that inequality in the distribution of rewards results from the inequalities in power is questionable. Conflict theorists can argue that in society at large one's position is determined by one's starting point. Children of the rich have considerably better life-chances than children of the poor. In science, position is far from inherited. Most of the men who occupy positions of power and influence have made substantial contributions to scientific knowledge. Although it is possible to show that those scientists who start out at the prestigious graduate schools have a head start, this does not mean that science is necessarily particular-

43. Ralph Dahrendorf, *Class and Class Conflict in Industrial Society* (Palo Alto, California: Stanford University Press, 1959), p. 165.

istic. If the brightest students go to the best graduate schools, the system would still be employing universalistic criteria.

In addition to asserting that the scientific rich wittingly or unwittingly employ power to favor their intellectual "children," conflict theorists would argue that the structure of science reflects the power structure of society at large.[44] Thus, the relative prestige of the various sciences would be determined by the extent to which they are needed by military and industrial power elites. Sciences with military or industrial applications receive more resources and have higher prestige than those with no military or industrial applications. Conflict theorists would also view the social relationships within science in terms of exploitation. Powerful scientists exploit graduate students, junior colleagues, and technicians in order to make their discoveries. The scientific "lower class," although essential for scientific progress, receives only scanty rewards. Certainly, in support of the conflict theorists, it cannot be denied that applied science is more heavily supported than pure science. However, the prestige of the various sciences probably has more to do with their state of intellectual development than their utility to industry and the military. Prestigious fields like high-energy physics currently have few if any practical applications. As to the extent to which social relations in science fit the exploitation model, we shall discuss this question in detail later in this chapter.

The core of conflict theorists' critique of the functional theory of stratification comes down not so much to questioning the rationale for having differential rewards attached to social positions, but to questioning the criteria upon which individuals are assigned to various positions. Conflict theorists would probably not argue that there should be no differentials in recognition accorded various scientists. But they might argue that the social processes through which individuals are singled out for recognition are not universalistic. This brings us directly back to one of our main topics of concern: the processes through which individuals are evaluated and located in the scientific stratification system.

THE ROLE OF UNIVERSALISM IN SCIENTIFIC STRATIFICATION

At the core of the institutional structure of science are a set of basic values and norms, which serve to orient and guide as well as control the behavior of scientists. The central values of universalism,

44. The following section has benefited from discussions of these issues with Professor Lewis Coser.

disinterestedness, organized skepticism, communism, rationality have been discussed in depth by Merton, Barber, Hagstrom, Storer, as well as by other sociologists of science.[45] The value of most critical importance to our analysis is universalism.

Universalism refers to a mode of characterizing social objects. Shall the object, which in science is usually the individual scientist or piece of scientific research, be judged on the basis of a universal frame of reference, or by particular attributes that the scientist possesses?[46] Particularism, the logical antithesis of universalism, is prototypically seen in nepotism. Given that science is ideally both rationally and universalistically organized, the criteria for judgment will be the quality of a scientist's research and his performance in roles which directly contribute to furthering scientific goals. Scientists should not be evaluated on any other criteria, such as the irrelevant ones of sex or race. The universalism-particularism dichotomy must be seen as a continuum. Most institutions are ideologically and rhetorically committed to universalistic principles; few, in fact, realize the ideal.

In general, there has been a trend toward greater universalism in nineteenth- and twentieth-century industrial society. Rationality and efficiency, long the hallmarks of capitalist business organization, continually interact with universalism. Blau and Duncan have noted that an increase in universalism has had profound implications for the stratification system. "The achieved statuses of a man, what he has accomplished in terms of some objective criteria, becomes more important than his ascribed statuses, who he is in the sense of what family he comes from. This does not mean that family background no longer influences careers. What it does imply is that superior status cannot any more be directly inherited but must be legitimated by actual achievements that are socially acknowledged."[47]

Blau and Duncan's observations, of course, are generalizations. It seems intuitively obvious that the replacement of particularistic with universalistic criteria of evaluation is a goal which is achieved

45. Merton, "Science and the Democratic Social Structure," in *Social Theory and Social Structure;* Warren Hagstrom, *The Scientific Community;* Norman Storer, *The Social System of Science*. See also the recent discussion by André Cournand and Harriet Zuckerman, "The Code of Science: Analysis and Reflections on Its Future," The Institute for the Study of Science in Human Affairs, Columbia University, 1970.

46. Talcott Parsons, *The Social System* (Glencoe, Illinois: The Free Press, 1964), especially pp. 101–12; Max Black, ed. *The Social Theories of Talcott Parsons* (Englewood Cliffs: Prentice-Hall, Inc. 1961).

47. Peter Blau and Otis Dudley Duncan, *The American Occupational Structure* (New York: John Wiley & Sons, 1967), pp. 429–30.

to varying degrees by different institutions. While economic organizations have moved in the direction of universalism with the decline of family capitalism,[48] we do not think of the business world as applying universalistic criteria of evaluation to the same degree as the professions or science.[49] Total universalism remains a "utopian" vision.

If evaluation in science is predominantly universalistic, there should be little, if any, "discrimination" in science. What do we mean by an absence of discrimination? Discrimination involves the use of functionally irrelevant statuses in evaluating an individual's performance as a scientist.[50] In the ideal social system, individuals would be evaluated on their competence at performing their roles. Position in the stratification system would be determined by performance. For the scientist, these roles include, among others, researcher, teacher, and administrator. If any scientist is evaluated as doing outstandingly well in these areas, he should be rewarded and occupy a prestigious and therefore desirable position in the hierarchy of the system. Individual characteristics such as race, age, sex, religion, nationality should not be relevant to the evaluation of him *qua* scientist. They are, as far as we can tell, functionally irrelevant characteristics for doing good science. To the extent that such extraneous variables are used to evaluate a scientist, we may say that "discrimination" is operating; that functionally irrelevant characteristics have been imported.

It is trite but hardly trivial to say that discrimination is widespread in most institutional spheres. One of the fundamental questions we attempt to answer in this book is: to what extent, if any, does discrimination exist in science? Or to put this question differently, to what degree is the stratification system of science dominated by universalism and achievement, as opposed to particularism and ascription? Does science really approximate the ideal of a meritocracy? Can science, in fact, be a viable institution without the predominance of universalism? Finally, under what conditions, if any, will nonuniversalistic criteria be employed?[51]

There is another set of problems generated by a consideration of

48. Daniel Bell, *The End of Ideology* (Glencoe, Illinois: The Free Press, 1960), chapter 2, "The Breakup of Family Capitalism."

49. Parsons, *The Social System*.

50. The concept of functionally irrelevant statuses was first introduced to us by Robert K. Merton in one of his courses and has not yet found its way into print.

51. See Barber, *Science and the Social Order*, for a detailed examination of Nazi science, where nonuniversalistic criteria were used to judge performance of scientists.

universalism in science. We must analyze the effects a universalistic reward system has on various levels of the stratification system. First, what are the consequences of universalism for the entire social system? Second, how are individual scientists affected by the presence or absence of universalistic standards of evaluation? We might immediately conclude that the adherence to strict universalistic criteria can only benefit both the system and the individual scientist. Yet, the problem is more complex. For example, how do scientists who are "failures" in the system rationalize or cope with this institutional judgment?[52] Third, to what extent are criteria of evaluation which are irrelevant from the point of view of the individual scientist, very much relevant to the institutional goals of the system? In other words, discrimination against individuals may result in overall benefits for the system. The question that is posed by this possibility is: how can one derive a calculus of benefits for the strain between responsibility to the individual and responsibility to the goals of the social system?

Among the various consequences of universalism in science would be the distribution of rewards solely on the basis of demonstrated ability. Clearly, then, the existence of a universalistic reward system does not imply egalitarianism. As the gaps in performance become wider, the stratification system would indeed become more differentiated in terms of recognition.

THE PROCESSES WHICH DETERMINE SCIENTIFIC STRATIFICATION

Having described the stratification system of science, we turn to an analysis of the social processes which produce the observed patterns. We consider a variety of factors which tend to produce differential rankings.

Scientific talent

Perhaps the most obvious variable influencing location in the stratification system is "innate" scientific ability. A plausible hypothesis

52. The problems associated with failure in modern society have become of increasing interest to scholars. We shall discuss this problem in relation to science, but consider one opinion on the problem as described by Michael Young in *The Rise of the Meritocracy* (Harmondsworth, England: Penguin, 1961): "Today all persons, however, humble, know that they have had every chance. They are tested again and again. If on one occasion, they are off-colour, they have a second a third and fourth opportunity to demonstrate their ability. But if they have been labelled 'dunce' repeatedly they cannot any longer pretend. . . . Are they not bound to recognize that they have inferior status . . . because they are inferior? . . . For the first time in human history the inferior man has no ready buttress for his self-regard" (pp. 107–8).

might hold that a scientist's success or failure is largely determined by the quality of his mind. And indeed evidence exists which in part substantiates this point of view. Certainly, successful scientists tend to score extremely high on standardized tests of intelligence. But the story is more complex than it first appears. Price, Harmon, and others report that the average I.Q. of physics Ph.D.'s is about 140.[53] These scientists are in the top 1 percent of the distribution of measured intelligence in the United States. But the fact that we are dealing with such an elite group of men to begin with, makes it difficult to assess in a measurable way the effect of native ability on success in science. In fact, Bayer and Folger have shown that for at least one sample of scientists there is an insignificant negative correlation ($r = -.05$) between I.Q. and scientific achievement (measured by citations).[54] This evidence suggests that although it requires a high I.Q. to become a scientist, once the Ph.D. is obtained differences in intelligence (as measured by I.Q.) do not influence scholarly productivity. Of course, these data do not consider the difficulty in measuring certain types of creative ability and imagination among men who, for all practical purposes, are "off the curve" in terms of measured intelligence. The fact that the I.Q. does not seem to be related to the quality of work produced strikes us as strange because we are continually identifying young scientists who are earmarked for greatness. Most scientists of the first rank maintain that it is usually easy to identify the young man with truly extraordinary native scientific talent. It may be that I.Q. tests are too broad to differentiate specific types of intellectual skills. Until we have better means of measuring such skills, we are left with the conclusion that creativity, or natural ability, constitutes a necessary but nonsufficient condition for success in science.

Data that we gathered on 499 academic physical, biological, and social scientists corroborate the Bayer and Folger findings on the correlation between I.Q. and citations to published research.[55] For this sample of university scientists the zero-order correlation between number of published papers and measured intelligence was $r = .05$; the correlation between quality of output, as measured by citations, and I.Q. was $r = .06$. While we found no correlation

53. Lindsey R. Harmon, *Scientific Manpower,* 1960 (NSF 61-34, May 1961), pp. 14–28, as cited in Price, *Little Science, Big Science,* p. 52.

54. Alan E. Bayer and John Folger, "Some Correlates of a Citation Measure of Productivity in Science," *Sociology of Education* 39 (Fall 1966).

55. The basis on which this sample of scientists was drawn is described in greater detail in chapter 5 and Appendix A.

between patterns of scientific output and I.Q., we did find a significant relationship between I.Q. and the prestige rank of the scientist's academic department ($r = .27$). Regression analysis suggests, furthermore, that high ability as indicated by I.Q. has an independent effect on this form of recognition. When we regressed rank of current department on I.Q., rank of doctoral department, prestige rank of the department at which the scientist held his first job, and scientific output, the partial regression coefficient between I.Q. and rank of current department was not reduced ($b^* = .26$). In short, these data offer preliminary evidence that although innate ability (as measured by I.Q.) is not correlated with the quality of research role-performance, it is in some way recognized. Several interpretations of these data are plausible.

Although it is usually quite easy for a scientist to distinguish between work of sharply divergent quality, it is not as easy to make judgments about work of similar quality. Scientists do not evaluate the quality of work by looking it up in the *SCI*. Thus, although I.Q. is not correlated with the number of citations to a scientist's work, it may be correlated with the perceptions that one's colleagues have of the quality of work. Let us consider how work of roughly equal quality produced by two scientists with differing I.Q.'s might be evaluated. If I.Q. measures to some extent verbal facility, then scientists who are more verbal may well be able to "sell" themselves to their colleagues better than others who are less so. There is a payoff in academic life for verbal ability, independently of other aspects of role-performance. Indeed, the proper "presentation of self," reflected in an ease with language and taken as an indicator of "intelligence," may account for the modest independent effect of I.Q. on the prestige of the academic department at which a scientist is located. Verbal scientists with high I.Q.'s may be able to convince their colleagues that their work is of higher quality than a completely objective evaluation might indicate.

Another interpretation for the correlation between I.Q. and prestige of academic department might be that "gatekeepers" within academic departments may recognize innate talent and choose to reward it—even if they recognize that the more talented scientist has not produced more or better research than the scientist who is seen as less intelligent. Scientists with high I.Q.'s may be deemed to have greater unrealized potential. They may also provide greater intellectual stimulation in interaction with their colleagues, and accordingly receive rewards for being "house intellectuals." A third interpretation of the finding would be that more able scientists, in terms of measured intelligence, may have a more accurate understanding of the operation of the reward system within academic departments, have more detailed knowledge of the normative system which operates in these departments, and have greater ability to control their immediate environment in such a way as to receive greater payoff for equal work. All these possibilities deserve further research.

When considering the effect of nonsociological variables on a scientist's place in the stratification system, attention also should be paid to the influence of such factors as energy and motivation. Assuming native ability among young scientists, a critical variable influencing success may be sheer work capacity. To be a successful scientist one must expend enormous amounts of energy, working long hours. Probably more important than physical stamina, however, is psychological motivation. Unlike in most other occupations, there are very few constraints on the professional behavior of scientists.[56] A university professor, for example, spends only a small proportion of his work week on formal instructional or administrative obligations. The rest of the time he is free to do what he wants. To spend a high proportion of this "free" time in self-directed research requires a good deal of self-discipline and motivation. The discipline demanded of the scientist is a strict one, requiring considerably more than putting in many hours. Thinking creatively is perhaps one of the most difficult types of work. The effort involved in pushing one's mind beyond its current limits is great. Because this work is difficult, scientists will find an unlimited number of ways to avoid it even while they are supposedly working. To be successful, a scientist must have the self-discipline to work long hours and to work productively. Such self-discipline and motivation probably explains at least as much variance on scientific success as native ability.

The psychological correlates of success in science have been discussed by Roe and Kubie, among others, but little is known about the ways in which personalities influence mobility in science.[57] Those studies that have been done suggest that firstborns, and adolescents with introspective, "inner-directed" personalities, tend to be more

56. This discussion applies only to academic scientists.

57. Anne Roe, *The Making of a Scientist* (New York: Dodd, Mead, 1952); Lawrence S. Kubie, "Some Unsolved Problems of the Scientific Career," *American Scientist* 42 (1954):104–12.

interested in science, and frequently do well once they enter science. Yet the methodology of these studies leaves so much to be desired as to render them of limited value. Aside from their inadequate samples and their failure to control for extraneous variables, they never consider the various ways in which personality interacts with social structure to produce patterns of behavior. Much research remains to be done in this area.

Once we move beyond the influence of native ability, motivation, and personality factors as determinants of success, we may examine the various aspects of the social structure of science for clues to the processes by which individuals find their level in the stratification system.

Accumulative advantage

In American science, a relatively small proportion of graduate schools produce a major portion of science Ph.D.'s; even fewer produce the scientists who will go on to receive recognition. During the post-World War II period, from 1950 to 1966, "four to five percent of the [doctorate-granting] institutions supplied 25 percent of the doctorates during the entire period; eleven percent of the institutions supplied 50 percent of the doctorates; and twenty-four to twenty-six percent supplied 75 percent of the doctorates."[58] These graduate departments are the most prestigious ones in the United States. Harriet Zuckerman has found that: "About half of all the Nobel laureates who studied in the United States received their degrees from just four universities, Harvard, Columbia, Berkeley and Princeton, which during the same period produced just 14 percent of all science doctorates."[59] Zuckerman notes that the same pattern holds for National Academy members. In 1969 73 percent of the 710 members of the National Academy who earned their Ph.D.'s in the United States received their degrees from just ten universities.[60] What these data point to is the relationship between early location in the stratification system and ultimate success.[61]

58. Publication 1489, *Doctorate Recipients from United States Universities: 1958–1966* (National Academy of Sciences, Washington, D.C., 1967).

59. Zuckerman, "Stratification in Science."

60. Don E. Kash et al., "University Affiliation and Recognition: National Academy of Sciences," *Science* 175 (10 March 1972):1076–84.

61. For further treatment of the influence of graduate education and social location on the destiny of the scientist in the stratification system, see Diana Crane, "The Academic Marketplace Revisited," *American Journal of Sociology* 75 (May 1970):953–64; Lowell Hargens and Warren Hagstrom, "Sponsored and Contest

Through processes of self-selection and social selection, most young scientists with potential attend the better graduate schools. Thus in one sense most scientists begin at the top. At an early age, they have access to some of the greatest minds in American science. The most talented graduate students have the opportunity to become working members of collaborative research teams early in their careers. Students at the top departments, judged to have outstanding potential, are kept on as faculty members, or they receive positions as junior faculty in other departments of the first rank. Those students who do not perform brilliantly (the overwhelming majority of any cadre of students, of course) are "assigned" to academic positions removed from the cutting edge of science. (There are also the few brilliant scientists who, for a variety of reasons, choose to move to less prestigious departments.) From this majority come most of the scientists who turn away from academic research; those who take jobs in industry, government, and teaching-oriented colleges. The young scientists who have been retained at top-ranked departments compete for postdoctoral fellowships, tenure positions, collaboration with world-famous scientists, and other scarce resources. Since most scientists receive their Ph.D.'s at the top graduate departments and because only the most successful receive tenure at these departments, the flow of talent in science is from the research centers to the outer fringes. It has been said that in the business world, people "rise to their level of incompetence." In science, people sink to their level of competence.[62]

Social and self-selection play a further role in the stratification process. Able young scientists seek out older eminent men to work with. Many of these young men in a hurry actually travel all over the world, often with little financial support, in their search for men with whom they can work and learn. An instructive example is provided by the experiences of James D. Watson, as he describes them in *The Double Helix*.[63] Not only did Watson seek and find Delbruck, but he travelled all over Europe in search of scientists who could

Mobility of American Academic Scientists," *Sociology of Education* 40 (Winter 1967):24–38; Lowell Hargens, "Patterns of Mobility of New Ph.D.'s among American Academic Institutions," *Sociology of Education* 42 (Winter 1969):18–37; Bernard Berelson, *Graduate Education in the United States* (New York: McGraw-Hill, 1960.)

62. William J. Goode, "The Protection of the Inept," *American Sociological Review* 32 (February 1967):5–18.

63. Watson, *The Double Helix*.

help him find clues to the mystery of DNA. His attachment to the Cavendish in England could not be said to be wholly unpragmatic. In the process, Watson had a growing subterranean reputation. On the basis of his work prior to the publication of that now famous paper in *Nature,* he was only a bright, aggressive young molecular biologist; yet many of the world's leading biologists were familiar with his name and knew vaguely what he was working on. Even before his pathfinding work, Watson could be said to be travelling in elite circles; his reputation truly preceded him to the ears of men like Pauling. It is, then, the informal structure of social relationships which gives the young scientist with potential but little proven talents the opportunity to work within the elite stratum of science.

This informal community of age-graded elites is an important aspect of the stratification system. Evidence suggests that men with talent who interact with established eminent scientists have a high probability of "making it." More than half of the Nobel laureates over the past ten to fifteen years have been students of earlier prize winners.[64] Moving from junior collaborator to senior author on important discoveries is not an uncommon process in contemporary science. We know little about how the careers of young scientists are affected by a lack of opportunity to collaborate with older, eminent scientists. It is difficult to ferret out the independent effect of individual capabilities and the effect of unequal research opportunities. A basic question remains: to what extent does a young scientist who initially does not have access to these opportunities find himself forever scientifically behind? How does the talent of this scientist get recognized? Does the system cut itself off from the young men who are the big fishes in the small ponds? Some recent data on mobility in science suggest that the process is not clear-cut. There are scientists moving into the elite departments from less distinguished ones; and an initial position in a top department does not ensure a life-time sinecure for those scientists who fail to produce.[65]

What we really are considering here are various aspects of the process of accumulative advantage. By virtue of being in top graduate departments and interacting with influential and brilliant scientists, some scientists have a social advantage in the process of stratification. Once position has been established in this initial phase, the probabilities may no longer be the same for two scientists

64. Zuckerman, "Nobel Prizes."
65. Hargens and Hagstrom, "Sponsored and Contest Mobility."

of equal abilities. The one who is strategically located in the stratification system may have a series of accumulating advantages over the one who is not a member of the elite corps. Indeed, these advantages may ultimately influence his position in the stratification system. This is not to say, of course, that he does not also face a series of tough tests along the way. The effect of structural variables in determining various forms of recognition will be a constant reference point in the empirical sections of this book.

There is another widely acknowledged advantage to being a member of a top research department or laboratory. Scientific resources and facilities are not equally assigned to research departments by those in charge of their distribution. On the contrary, there is a skewed allocation of these facilities; for the most part, prestigious research centers get a vastly disproportionate share of the wealth. The consequence for the individual scientist in these centers is a clear advantage in the production of new ideas and new publications. Of course, increased publications place these scientists in a better position for receiving future funds. The generic problem behind the concept of accumulative advantage is whether this process is an outcome of the unequal distribution of talent, which tends to cluster at the prestigious centers, or whether "talent" is a result of the unequal distribution of resources and facilities. In either event, the process of accumulative advantage contributes to the sharpening of stratification lines.

Published scientific output

Although the process of accumulative advantage helps some scientists to get started and to maintain their superior records, if science is truly universalistic and rational a man's location in the stratification system will ultimately depend on his published scientific contributions. We have already noted that science is highly stratified in terms of publication rates. A relatively small proportion of those active in research publish almost half of the total scientific literature. Since making new discoveries is a primary goal of the scientist, rewards would, logically, closely parallel the distribution of publications. But consideration must be given, of course, to the all-important distinction between the quantity and the quality of research publications. Equating quantity with quality will not do as a basis for rewards. In a stratification system based on universalism, high quality work will be rewarded, regardless of its source. Recognition for outstanding research is the pillar which supports the entire scientific society. Without just rewards for research well done, science would probably deteriorate. Furthermore, unless there is recognition of excellence, the bases for assigning individuals to different positions in the system will seem totally arbitrary.

There are some scientists who hold that differential productivity rates in science are a result of the attitudes and "biases" of journal editors and referees. The scientists who adhere to this conspiracy theory see the leading journals as being controlled by a small circle of men at a few major universities. The editors, it is believed, are more likely to accept papers for publication that are produced by members of the most prestigious departments. Consequently, publication rates are seen in part as the result of a discriminatory process. Zuckerman and Merton have recently collected data which cast doubt on the validity of this theory.[66] Although rejection rates differ for the various scientific disciplines, virtually all of the journals in the hard sciences accept more than 50 percent of all papers that are submitted to them. The world's leading journal, the *Physical Review,* accepts about 85 percent of all the papers submitted to it for evaluation. The figures are similar for all the hard sciences: 76 percent of physics papers, 69 percent of those in chemistry, and 71 percent in biology are accepted for publication. In contrast, the proportion of articles submitted to social science journals that are eventually rejected is far higher. More than 80 percent of the papers submitted to the *American Sociological Review* are rejected.[67] With such a low rejection rate in the physical sciences, it hardly seems likely that editors are systematically rejecting papers on the basis of their source. While the rejection rate of papers produced by scientists located at less distinguished departments is higher than for those produced by members of top departments, still a clear majority of papers which come from less prestigious universities find their way into print. Furthermore, the small difference in rejection rates could just as easily be a result of differences in quality as of discrimination. Before discrimination can be proven it must be shown that for papers of *equal quality* those written by scientists at prestigious departments have a higher probability of being accepted. Since papers rejected by one journal can be submitted to another, we can at least tentatively conclude that in the hard sciences almost any paper can be published.

We may draw two conclusions from these data. First, there

66. Zuckerman and Merton, "Patterns of Evaluation."
67. Ibid, p. 52.

appear to be informal norms which determine whether a paper is worth submitting to a journal. At the same time, the journals are attempting to publish most materials which in any sense of the term could be considered contributions. Second, there does not seem to exist in modern science any differential access to publications. The journals are open virtually to all legitimate scientists, and different productivity rates probably cannot be attributed to discriminatory practices. The fact that scientists of disparate backgrounds have equal access to the journals does not necessarily imply that the community will treat their work equally. Nor does it explain the reaction of the community to different career publication patterns—a response that will largely determine a scientist's ultimate position in the occupational and prestige hierarchies of the social structure. A detailed discussion based on empirical data for patterns of publications and the scientific community's response to them will be presented in chapter 4.

CONSENSUS AND AUTHORITY IN SCIENCE

A universalistic evaluation system is one in which the same criteria are employed in judging all individuals. Science is universalistic to the extent that all scientists are evaluated according to the quality of their scientific contributions. This does not mean (even if science were a real meritocracy) that there would be no conflict within science as to what is high-quality scientific work. If science were a utopian institution, there would exist virtual unanimity as to what constituted high-quality contributions. Although science is clearly not such an institution, there does appear to be a relatively high level of consensus as to what constitutes outstanding work, what are important problems to be addressed, and what are acceptable empirical techniques for testing scientific theories.

How is intellectual consensus maintained in science? According to the classical view of the history and philosophy of science, consensus is determined by the empirical phenomena themselves. Theories supported by empirical observation would become part of the consensus; theories at odds with observable "facts" would be discarded. Recent work in the history and philosophy of science by such men as Kuhn and Polanyi have made us accutely aware of the fact that scientific truth is transitory. The basic message of Kuhn is that consensus determines truth, not truth consensus. There are no eternal truths, only a series of shifting consensuses as to how the world should be viewed. Once we accept the notion that consensus does not automatically spring from nature, we are forced to pay more attention to the sociological processes through which consensus is developed, maintained, and eventually shifted.

One of the primary mechanisms through which consensus is maintained is the practice of vesting authority in elites. Indeed, the maintenance of consensus is an additional function of scientific stars. When a new idea is proposed, it must be evaluated: should the new idea become part of the consensus or should it be discarded? In the process of evaluation some opinions count more than others. The rationale for vesting intellectual authority in elites has been succinctly expressed by Ziman:

> Science is a social system in which leadership and intellectual initiative naturally gravitates to the individuals who can exercise it. It is commonplace in the discussion of the concept of "equality" in ordinary society, to remark that all men are not born as brilliant as Einstein, and therefore there is no reason why all men should be treated the same, etc. etc. The fact is that all theoretical physicists are not born as brilliant as Einstein, either, nor all biophysicists as smart as Crick, nor all radio astronomers as inventive as Ryle, and so on. The actual talents of those actively engaged in modern science spread over an enormous range, from the genius with 100 brilliant discoveries to his credit to the dullard with a Ph.D. and one uninteresting published paper in an obscure journal. One would not be human if one did not listen rather more carefully to the former than the latter, nor if one did not seek to be heard at least in proportion to one's own estimate of one's own standing on this scale.[68]

Scientific progress is in part dependent upon maintaining consensus by vesting intellectual authority in stars. Without consensus, scientists would go off in hundreds of different directions, and science might lose its cumulative character. The stars in a particular field determine which ideas are acceptable and which are not. Abandoning the principle of authority would eliminate a rational basis for discarding poor or irrelevant work. As can be seen in the following remarks of Polanyi, the authority principle lies at the heart of the evaluation system of science—indeed, at the core of the social structure of science:

> The continued existence of science is an expression of the fact that scientists are agreed in accepting one tradition, and that all trust each other to be informed by this tradition. Suppose

68. Ziman, *Public Knowledge*, pp. 31–32.

scientists were in the habit of regarding most of their fellows as cranks or charlatans. Fruitful discussion between them would become impossible and they would no more rely on each other's results nor act on each other's opinion. Thus their mutual collaboration on which scientific progress depends would be cut off. The processes of publication, of compiling textbooks, of teaching juniors, of making appointments, and establishing new scientific institutions, would henceforth depend on the mere chance of who happened to make the decision. It would become impossible to recognize any statement as a scientific proposition or to describe any one as a scientist. Science would become practically extinct.[69]

The stars in a particular field determine which ideas are acceptable and which are not. Sometimes, because of this, new ideas are resisted or ignored; but as Polanyi has argued, the positive consequences of maintaining consensus probably outweigh the negative consequences of occasional failures in the evaluation system.

Polanyi relates a relevant story from his own experience as a scientist.[70] As a young chemist, he proposed an unorthodox theory of adsorption which went contrary to the prevailing ideas held by Einstein and Bohr. This theory, which contradicted existing premises, was rejected but later turned out to be correct and was incorporated into the general fund of knowledge. What makes this case of particular interest is Polanyi's public argument that it was right for his ideas to be resisted. Polanyi claims that as a youngster and without scientific reputation he had not earned the "right" to have ideas so contrary to prevailing theory accepted without initial resistance. His theory turned out in retrospect to be correct. But if all such unorthodox theories were to be given credence, science would be harmed since most of them would turn out to be incorrect. If scientists were to accept every unorthodox theory, the established consensus would be destroyed and the intellectual structure of science would become chaotic. Young scientists would be faced with a multitude of conflicting and unorganized theories and would not have any guides in doing their research.

Given the role that stars play in exercising authority, establishing and maintaining consensus, serving as gatekeepers for scarce resources and as referees for journals, consider what would happen to

69. Michael Polanyi, *Science, Faith and Society* (Chicago: The University of Chicago Press, 1964), pp. 52–53.

70. Michael Polanyi, "The Potential Theory of Adsorption," *Science* 141 (1963):1010–13.

a field without stars. It is unlikely that a modern science could function at all. Scientists must be found to fill these important positions. If the positions are filled by "average" scientists, it will be difficult for the authority exercised to be granted legitimacy. It is only when the scientific community sees those exercising authority as deserving of it that the authority will be accepted. Thus, the sociological process of stratification through which stars are created is probably necessary for the maintenance of consensus and the orderly functioning of the scientific community.

The image of consensus may well lie behind the belief held by many scientists that, even if science is stratified, the frequent concomitants of stratification in other institutions—the competitions, jealousies, and inequalities—are, for some reason, relatively absent from science. The very definition of consensus implies some form of harmony, cooperation, and group solidarity. What appears to mislead some is the assumption that the existence of a relatively high level of consensus in the pure sciences on acceptable methodologies, theories, and criteria of good work, is incompatible with social inequality. On the contrary, the two are quite compatible. It is quite possible to have consensual agreement and sharp inequalities in power, authority, and rewards. In fact, it can be argued that the elements of consensual agreement in science reinforce sharp stratification. If there is agreement on what constitutes outstanding research, then it becomes considerably easier to institute a legitimated reward system that honors research which meets the clear standards of excellence. Correlatively, it is apparent why some scientists reach the top and others go unrecognized.

Thus far we have not discussed the distribution of power in science. Is there, in fact, a meaningful distinction to be made between power and authority in the scientific community? Is there a "power elite" or a "veto group" in science? Before we can discuss this question we must consider definitions of the two concepts. Weber defined power as "the probability that one actor within a social relationship will be in a position to carry out his own will despite resistance, regardless of the basis on which this probability rests." [71] By contrast, authority, as defined by Barnard and others, rests on acceptance by subordinates of the legitimacy of the superordinates' right to give orders. The essence of authority and its effective operation lies in its acceptance by the governed. In science,

71. Max Weber, *The Theory of Social and Economic Organization* (Glencoe, Illinois: The Free Press, 1947), p. 152.

authority is the socially legitimated right to decide what is important and what is not. The exercise of authority influences only indirectly the scientists subject to it. Thus when a scientist's work is highly evaluated it does not directly propel him into a prestigious position. Likewise, when a negative evaluation is given to work it does not directly lead to downward mobility. Power in science would be the direct influence of one man or a group of men on another's career. When a scientist is denied tenure or promoted, given a research grant or denied access to research facilities, given a fellowship or denied one, power is being exercised. Since there is a relatively close link between the evaluations made of a scientist's work and the progress of his career, the gap between authority and power is small.

In science there are definitely some men in structural positions who can strongly, if not ultimately, influence the fate of other men's work and careers. We are not dealing here with the authority of an Einstein or of adherents to the Einsteinian conceptual paradigm to determine what is a legitimate or acceptable piece of science. We are dealing with the ability of one scientist to affect another's career. Which groups and individuals hold positions of power in science? There is substantial overlap in the groups having power and those having prestige. The two groups having the capability of wielding power are the same two groups which make up the prestige elite: those scientists who have earned recognition for their outstanding contributions to knowledge and those who hold key administrative positions. They comprise the relatively small number of scientists who largely control mobility within science. Those in the first group, as members of the most prestigious departments, decide who among the younger scientists can receive continuing appointments at the major institutions; they nominate new men to positions and membership in honorary societies; as editors of journals they decide what should and should not be published. Those in the second group, the administrators of foundations, laboratories, and government agencies, are influential in setting policy on what sciences or scientific specialties should be heavily funded. They can determine what specific research areas are to receive priority, and what individuals are to receive support for their research programs. In short, they largely control the resources and facilities of modern science, while playing a central role in determining the foci of scientific attention. The significance of their positions is underscored when it is realized that these administrators are largely responsible for allocating some 1.5 billion dollars annually for research.

At first it seems as if there are no courts of appeal after decisions have been reached by those in positions of power. Their judgments of the ability of men applying for positions or for research grants appear to be final. In the scientific community power over the many seems to lie in the hands of the few. But one important modifying point must be made. One organizational feature of power in American science is its relative diffuseness. Although power is concentrated in the hands of relatively few scientists, they are not located in any centralized agency. In other words, there are a variety of foci of power in different universities, journals, scientific societies, and government and private funding agencies. Unlike the situation in some European countries, there really is no single organization that can determine the fate of a scientist and his work. If one university department does not offer a scientist tenure, this does not necessarily jeopardize his chances of receiving tenure at another university of equal rank.

How does science administer the transfer of power from one generation to another; more important, what mechanisms operate to shift power and influence from adherents of an older paradigm to adherents of a newer one? Clearly, scientific revolutions do not take place overnight. There is a gradual replacement of older elites with younger men who may hold new ideas about the conceptual framework of their discipline. Max Planck once observed: "A new scientific truth does not triumph by convincing its opponents and making them see the light, but rather because its opponents eventually die, and a new generation grows up that is familiar with it." [72] Nonetheless, continual change is possible in science because, for the most part, fundamental discoveries which may significantly alter the conceptual apparatus of a science are generally not perceived by men in power as undermining their positions. In fact, prestige is often conferred upon scientists and organizations that support pioneering research. Further, there is no direct inheritance of power, privilege, or wealth in science. A scientist's intellectual heirs, his students, must achieve power on their own. Since science is fundamentally committed to intellectual change, the most powerful scientists do not hesitate to train students even though they know that these students may eventually produce work which will make their own obsolete.

The basic problem in analyzing power in science is that its

72. Quoted in Barber, "Resistance."

exercise is so closely linked to legitimated authority. It is difficult to find instances in which power is exercised independently of authority. We shall present data in the following chapters which indicate the extent to which the "gatekeepers" and "status judges" in science, those scientists with the power to significantly influence career patterns through their evaluations of competence, base their evaluations on rational and universalistic criteria. If power is not abused in science; if it does in fact merge with authority, its application may not produce a sense of injustice at the social organization of science. It is indeed possible that the distinction between power and authority in science may not be a particularly useful one.

SOCIAL CONFLICT IN SCIENCE

Before we proceed to the data on the processes of stratification, one basic question must be addressed. Given the great amount of inequality in science, how much conflict exists between the various strata in contemporary science? A central theme in modern societies of widely varying social and political structures is the struggle to reduce salient inequalities between social groups.[73] To pose proper questions, we must begin with a tentative answer. Available evidence suggests that there does not exist significant "class" conflict in science between the "haves" and the "have-nots." In fact, in science it is difficult to find the rhetoric of social class distinctions. Nowhere in the literature produced by scientists can we find the Marxian conception that scientists who remain unrecognized constitute a class "in itself," if not "for itself." One rarely finds in the official meetings of science associations, or in the journals, discussion of basic *social* conflict in science. Indeed, its absence seems to be taken for granted. Of course, there is routinized intellectual controversy which goes on continually at all levels of science. Often there are heated and extended debates. Yet there is very little basic questioning of the legitimacy of the social structure of science by any identifiable group or stratum. There are sporadic cries that rewards are unfairly meted out to individuals, but no organized movement of protest exists against the structure of the reward system or the basis on which the performance of scientists is judged. This relative lack of social conflict is a sociological

73. For an interesting treatment of this aspect of the stratification literature, see "stratification" by Arthur Stinchcombe in the *International Encyclopedia of the Social Sciences* (New York: Macmillan and Free Press, 1968), vol. 15, pp. 325–32.

anomaly. In recent years, there have been increasing complaints from some quarters of the community that women scientists are unfairly treated in the allocation of rewards—in short, that there is sexual discrimination in science. Some data are presented in chapter 5 which test this assertion.

In order to understand why there is so little social conflict in science we must consider the relationship between the degree of inequality that exists and the various bases for solidarity between the various strata. Why do scientists who are differentially rewarded by their community nonetheless maintain a communal identity? Although there are presently no systematic empirical treatments of these questions, we can suggest several complementary factors which in part explain the preservation of communal solidarity in science.

One of the fundamental generalized beliefs in science is that every scientist in his own way contributes to scientific advance. This belief acts to prevent the emergence of conflict between the social strata. Rarely will we find scientists taking public issue with this position, even when they privately do not believe that it is accurate. The meaning of these remarks and the consequences of this ideology are manifold. It tells scientists that in the broad perspective of the history of science their work does count. They may not be highly rewarded; they may not receive prizes, public honor, or positions in prestigious scientific organizations; but their work is needed. This ideology is somewhat akin to a physiological construct which views a body as effectively functioning only when the thousands of cells that make up the various organs work in harmony. A breakdown anywhere in the system, top or bottom, can only result in the disorganization of the entire structure.

Obviously, the belief that one's work, whatever its form, contributes to the advance of science, must be a soothing thought to many relatively unproductive and noncreative scientists. Since most scientists publish very little in their lifetimes and since much of what they do publish is virtually invisible, this belief in the value of their contributions at least gives them a feeling of common undertaking and harmony of interests with their more successful colleagues. Let us not carry this argument too far. Scientists do hunger for recognition; but when it is not forthcoming they can fall back, to some extent, on the notion that their work ultimately does make some difference—that it is socially useful. Many scientists, in fact, articulate the belief that their work essentially involves filling in the details of the masters' designs, while others feel that they are working on the materials that the truly great artists of science will use in constructing the new scientific "revolution." This generalized belief does not only affect the relatively unsuccessful researcher, it also affects scientists who have turned their energies to teaching and administrative work. For these roles can also be said to be essential for the future development of science, and the men who fill them can find satisfaction in the hope that their work in some small way will help science solve some of the mysteries of nature.

We are suggesting, then, that at least one ideological feature of modern science acts, whether consciously intended or not, as a means of social control and constraint on organized "deviant" behavior. The virtual universal acceptance of this sense of common purpose has resulted in what Durkheim referred to as the "conscience collective." This notion of the harmony of interests for all scientists, regardless of their social position or the rewards that they have received, assumes the characteristics of a social fact. It is external to any scientist and it is constraining. However, it is not constraining in the sense of guilt; rather it takes the form of a deeply felt general belief in the common purpose of doing science.

One indirect indicator of this sense of identity can be found in the institutional norm of humility in science. Merton has discussed this norm extensively, particularly focusing on its relationship to the value placed by scientists on originality. Following Merton, humility takes the form of "acknowledging the heavy indebtedness to the legacy of knowledge bequeathed by predecessors," and of the insistence by the scientist on his personal limitations.[74] Newton's well-known epigram best expresses this value: "If I have seen farther, it is by standing on the shoulders of giants." We slightly alter the perspective on this norm. The norm suggests the truism that science is greater than any individual and that without the cumulative foundation of scientific knowledge no individual could hope to go far in building the edifice himself. But the norm functions in a different way as well. Humility counteracts the personal identification of ideas with men. Scientific ideas, once created, immediately become communal property, open to any and all. While the creator of the idea may remain associated with the construct, as in the case of eponymous recognition ("Boyle's law," "Planck's constant"), the ideas become part of a much grander system which is

74. Robert K. Merton, "Priorities in Scientific Discovery," pp. 463–64.

not associated with any one scientist. This norm accentuates and underscores the belief that science is the product of a communal effort, and lends added credibility to the idea that all members of the scientific community who act in accord with the prescriptions of scientific norms, in either an instrumental or expressive way, contribute to the welfare of the community as well as to scientific advance. Clearly, such generalized beliefs, if actually internalized by scientists throughout the social system, could go far to nipping in the bud any social conflict between scientists located at different levels in the stratification system.

What remains totally unresolved is whether this generalized belief is consistent with social actuality. Do most scientists, regardless of their "station," contribute to scientific advance; are they influential, either directly or indirectly, through their research or teaching, in moving science forward? This query is translatable into a series of empirically testable propositions. We suggest that there may have been some resistance among scientists to systematically addressing this question for reasons similar to those enumerated by Merton in his discussion of resistance to the study of multiple discoveries.[75] There is quite a bit at stake in such an investigation. If it should turn out that empirical evidence indicates that only a relatively small proportion of scientists have anything to do with progress in science, the consequences for the solidarity between social strata in science could be significant. Later in this book we take a first, tentative step toward empirically analyzing the authenticity of the generalized belief that scientists at all levels of the stratification system play a meaningful role in the development of scientific knowledge.

Socialization also tends to reduce the extent and intensity of social conflict in science. We have noted that a high proportion of scientists receive their training at the high-ranking departments. During their graduate careers they internalized the values and beliefs of those leading scientists who were their teachers. The master-apprentice relationship, which is a fundamental aspect of graduate science instruction, goes far toward inculcating youngsters with a set of values as well as knowledge. Values and attitudes are transmitted from teachers to students—those destined to occupy low-prestige positions as well as those who become highly successful. Further, the textbooks and monographs from which students "learn science" tend implicitly to reinforce a set of values, one of which is

75. See Robert K. Merton, "Resistance to the Systematic Study of Multiple Discoveries in Science," *European Journal of Sociology* 4 (1963):237–82.

that science is an identifiable community with a high value placed on the notion of common purpose.[76] A consequence of the socialization process may be that scientists who move away from the centers of research activity nonetheless maintain an identity with the goals and men who are at the frontier. This is one way in which ideology is transferred from one generation of scientists to another. As a result, scientists at lower levels in the stratification system may never develop a sense that their interests are fundamentally in opposition to those of their more eminent colleagues.

One of the striking features of intellectual conflict in science is that it rarely degenerates into polarized social conflict. Scientists rarely use the social characteristics of an author of a discovery as evidence against its validity. Indeed, the absence of ad hominem arguments within science sets it distinctly apart from its sister social sciences, where intellectual disputes often are transformed into social conflict.[77] Again Polanyi provides us with a compelling, if a priori, interpretation for this pattern of behavior. He notes that the fundamental basis of cohesion in the scientific community can best be seen, paradoxically, in the cases of intellectual conflict.[78] No matter how revolutionary the ideas of a "renegade" scientist may be, he bases his argument on "scientific tradition." The resisters of the new ideas also invoke the same tradition, which the heretic claims must be reconsidered. The scientist who produced the contrary ideas invokes memories of Pasteur, Lister, and others, whose ideas were initially resisted. But the important point is that all of the disputants are playing by the same rules. As Polanyi tells us, "even in the most profound divisions that have yet occurred in science, the rebels and conservatives alike remained rooted in the same grounds. Accordingly, these conflicts have always been settled after a comparatively short time in a fashion which has proved acceptable to all scientists." [79] In other institutional spheres, conflict over the economic and political organization of a society is based upon the invocation of fundamentally distinct intellectual and social

76. For a discussion of the socialization processes that lead to these commitments, see Thomas Kuhn, "The Essential Tension: Tradition and Innovation in Scientific Research," in Calvin W. Taylor and Frank Barron, eds., *Scientific Creativity: Its Recognition and Development* (New York: John Wiley and Sons, 1963), pp. 341–354; Warren Hagstrom, *The Scientific Community*, chapter 1.

77. Robert K. Merton, "Social Conflict over Styles of Sociological Work," *Transactions of the Fourth World Congress of Sociology* (1959), 3:21–46.

78. Michael Polanyi, *Science, Faith and Society* (Chicago: University of Chicago Press, 1964), pp. 50–53.

79. Ibid., p. 52.

traditions. This is one reason why conflict in these institutions seems to produce divisions that do not seem reconcilable by reexamination of any single tradition or agreed upon set of facts. For a Marxist and a non-Marxist, the basic premises of argument seem totally at variance. Consequently, there can be little, if any, agreement on what constitutes a fact, or a proof. In science, the latent structure of disputes is generally a common set of assumptions and standards of proof. Thus incipient intellectual controversies can be settled rapidly before they evolve into intense social conflict.

This leads directly to our final point. One possible reason why there is less social conflict between the strata in science than in other social institutions is that, contrary to what the conflict theorists would have us believe, there may be relatively little exploitation occurring in science. The one place where science seems to most closely parallel industrial organization is the large research team working in a huge laboratory. Much experimental high-energy physics is conducted in such a setting. On some experiments as many as a dozen or more scientists, assisted by graduate students and technicians, work together. To what extent does exploitation occur in such settings? Exploitation would occur to the extent that people making important contributions to the experiment did not get recognition. Here there is a crucial difference between the products of science and the products of industry. In industry the products are material goods with cash value. The value of a contribution to the product may be measured in terms of value added. In science the products are original ideas. The value system of science stresses the importance of originality. Of two scientists working equally hard on separate experiments, we would probably agree that the one making an original contribution deserves more recognition than the one publishing a pedestrian footnote. If we accept the premise that, in science, recognition should be distributed for contributions to advance in knowledge, then we can conclude that the technicians and graduate students working on an experiment do not deserve as much recognition as the scientists originating the experiment, analyzing the data, and writing the paper.

Are graduate students or junior colleagues denied their fair share of recognition even where they do make original contributions to an experiment? There are, no doubt, cases where this happens; however, as Harriet Zuckerman has pointed out, most eminent scientists who lead research teams make special efforts to see that their students and junior colleagues get the proper credit.[80] Sometimes leaders of a research team will even leave their names off papers to insure that credit is not misallocated. Of course, it is possible that despite the attempts of research leaders to be equitable the audience of the papers may misallocate credit. Merton calls this "The Matthew Effect." In chapter 7 we examine data which cast light on the extent to which such misallocation actually occurs in physics. Even if the audience does misallocate credit, it would be difficult to view this as exploitation. It would be equivalent to saying that the consumers rather than the capitalists are the exploiters of the proletariat.

In this chapter we have shown that the institution of science is a highly stratified one. There is at least as much inequality in science as there is in other social institutions. At the same time, we have indicated that the social processes that produce stratification in science may approximate the ideal of a meritocracy based on the application of universalistic principles, and may do this far more than most, perhaps more than all, institutions in our society. Implicitly we have suggested the existence of a significant paradox: that a relatively stable system which is full of sharp stratification of rewards is at its core fundamentally committed to change—intellectual change and "progress" as well as social change. Elites hold their positions only as long as they are active scientists. The notion of passing on "privilege" by anything short of scientific competence is foreign to this institution. Leading scientists can come from any social background if they have the stuff to produce good science. It appears possible that in science, from one generation to another, "the last may be first."

The story of stratification in science can only be understood in light of a detailed empirical examination of the social processes and structures that produce it. This task we shall now undertake.

80. Harriet A. Zuckerman, "Patterns of Name Ordering among Authors of Scientific Papers: A Study of Social Symbolism and Its Ambiguity," *American Journal of Sociology* 74 (November 1968):276–91.

4 LOCATION IN THE STRATIFICATION SYSTEM AND SCIENTIFIC OUTPUT

One of the major questions that must be considered in the analysis of any stratification system is how individuals are ranked. In studying basic societal stratification we might analyze the social processes through which individuals end up in the upper, middle, or lower classes. In this chapter we investigate the criteria employed and the processes involved in the social distribution of scientists to the different levels of the scientific stratification system. How does a particular scientist become a member of the scientific elite while another scientist remains obscure? We will be especially interested in the extent to which the reward system is universalistic and rational.

As we pointed out in the previous chapter, a scientific reward system is universalistic to the extent that quality of performance is the sole determinant of rank in the scientific community. The basic scientific roles are those of researcher, teacher, and administrator. Of these three, the value system of science emphasizes research; contributing to the advance of knowledge is the primary function of the scientist. If science does indeed have a universalistic reward system, the quality of published work should be the most important criterion employed in the distribution of rewards.

A reward system is universalistic if the same criteria are employed in evaluating all members. Another important attribute of reward systems is the extent to which they are rational. A rational reward system is one in which the universalistically applied criteria of evaluation are such that they aid the system in achieving its goals. In science, a rational criterion of evaluation would be the quality of research produced. There has been a long-standing belief that scientific reward systems are dominated by the "publish or perish" philosophy. If you want to get ahead you must churn out an endless stream of articles and books. It is sometimes said that the emphasis is on quantity of publication rather than quality.

The extent to which quantity or quality of research influences success may have a crucial impact on the rate of advance in a scientific discipline. The most important scientific problems are usually those which are difficult to solve. If success in science depends as much on the quantity of publications as on their quality, then young scientists may not be motivated to work on the important but difficult problems. Some scientists would hesitate to spend several years working on a difficult problem, the outcome of which is in doubt, when they could work on a series of simple problems that would be more likely to lead to quick publication. To what extent then does the quantity and quality of scientific research influence success in physics? To what extent is the reward system of physics universalistic and rational?

The results in this chapter are based primarily upon an analysis of the careers and scientific output of 120 university physicists, chosen from a sampling frame in which the population of university physicists was stratified along four dimensions: age, prestige rank of university department, productivity, and number of honorific awards.[1] A second set of data was collected to find out the relative prestige of awards received by members of the sample and the extent to which the 120 physicists are known within the national community of physicists.[2] Since two of the key variables in the analysis are the quantity and quality of research published by the sample of 120 physicists, we must have reliable measures of both. As a measure of the quantity of scientific output, we took the total number of scientific papers by each physicist as listed in *Science Abstracts*.[3] For our measure of the quality of scientific output, we have used material collected from the *Science Citation Index*. We used as an indicator of quality the total number of citations to a physicist's research in 1961, 1964, and 1965.

In our analysis we deal with four ideal types (see fig. 1) roughly described by the production of scientific papers and their quality (as assessed by citations). Type I is the *prolific physicist,* in the dual sense of producing an abundance of papers which tend also

1. For a detailed discussion of this sample and the way it was selected, see Appendix A.

2. A discussion of these data and how they were gathered also appears in Appendix A.

3. Two aspects of this decision should be noted. First, only papers and not books are included in research output, since physicists almost invariably publish their original research in papers, unlike in the humanities and the social sciences. Second, we shall be using the total scientific output (the cumulative number of papers published by each physicist) rather than productivity rates (average number of papers per year). We find that both measures exhibit the same patterns of relation to other variables examined in this chapter.

to be fruitful (that is, often used by others in the field). At the other extreme, Type IV, is the relatively *silent physicist:* he produces comparatively few papers and, judging from the paucity of citations to them, they do not matter much to the field of physics. The other two types remind us that the sheer quantity of published papers is not always correlated with their quality. Type II is the undiscriminating *mass producer:* the person who publishes a relatively large number of papers of little consequence. As a type, this person

Figure 1. Four types of academic physicists based on quantity and quality of published research

	Quality	
Quantity	High	Low
High	Type I Prolific	Type II Mass producers
Low	Type III Perfectionists	Type IV Silent

seems geared to getting many papers into print without much regard for their scientific significance. And finally, there is Type III who might be described as the *perfectionist:* the physicist who publishes comparatively little but in what he publishes has a considerable impact on the field. This type may include physicists who elect not to publish work which, in their own (possibly mistaken) judgment, does not measure up to sufficiently high standards. As a result, they are not the prolific researchers of Type I. The two variables used in constructing this typology, quantity and quality of research, are fairly highly correlated ($r = .60$). Thus the majority of physicists in this sample would be classified as being either "prolific" or "silent."[4]

4. These four types are only crudely approximated by the particular data used here, since the sample overrepresents those who are eminent. In fact, 45 percent of this sample had published at least thirty papers during their careers; and 52 percent produced a body of research which received 60 or more citations in the 1965 *SCI*. A much larger sample would be needed to identify the extreme types by more precise criteria, for example, the silent physicists of Type IV who had published no more than, say, two or three papers all told, and the prolific ones of Type I, with, say, one hundred papers or more. The same can be said about number of citations. Fifty-one per cent of the sample of 120 had more than five times the average number of five citations. Furthermore, a larger sample would enable us to identify intermediate and transitional types.

The dependent variable in this study is rank in the stratification system, or success. There are two interrelated components of scientific success. The first is reputational. The scientific elite are widely known among their colleagues and their work is highly regarded. In this study we have three measures of reputational success. Both the number of awards a scientist has received and the prestige of his highest award are indirect measures of scientific repute. Certainly, among the scientists who have the widest and best reputations will be found many recipients of prestigious awards. The third and more direct measure of reputational success is the scientist's visibility.

The second component of scientific success is positional. To be successful is to occupy a prestigious position in the scientific community. As a measure of positional success we use the prestige of the scientist's academic department. We shall first analyze the correlates of receipt of prestigious awards.

RECOGNITION THROUGH FORMAL AWARDS

Both the quality and quantity of published research have relatively strong correlations with the number of honorific awards a scientist has received.[5] The zero-order correlation coefficient between number of awards and production of papers is .45; the zero-order correlation between number of citations received and number of awards is .57. We are, of course, interested in the independent effect of quantity and quality on awards. We can assess this by computing a multiple regression equation and examining the standardized partial regression coefficients. When we do this, we find that quality of work has a far stronger influence on number of awards than does the sheer number of published papers. The standardized partial regression coefficient for the quality of work is .46; the partial regression coefficient for quantity is .18. If we decompose the multiple correlation ($r/R = .58$) we can ascertain the portion of the explained variance on awards that is uniquely due to each of the independent variables and the proportion due to their joint effect. About two-thirds of the explained variance on awards is due to quality of work alone; about one-third to the joint influence of quality and quantity of output. Sheer bulk of output has virtually no unique effect on predicting the awards a scientist has received. The zero-order correlation between quantity and awards results primarily from the fact that quantity and quality are correlated.

5. Number of awards and the prestige of highest award are correlated $r = .70$.

We may conclude that it is the quality of research rather than its sheer amount that is most often recognized through honorific awards.[6] Although they have published fewer papers than the prolific Type I physicists, the Type III ("perfectionist") physicists are just as apt to be accorded recognition, and both these quality-producers are far more likely to receive awards than the Type II mass producers who publish large numbers of papers indiscriminately.[7]

Decomposition of Explained Variance on Awards

	Quantity	Quality	Quantity + Quality
$R^2_{N(QP)} =$	p_{NP}^2	$+ p_{NQ}^2$	$+ 2\, p_{NQ}\, p_{NP}\, r_{PQ}$
$.58^2 =$	$.18^2$	$+ .46^2$	$+ 2\,(.18)\,(.46)\,(.60)$
$.3364 =$	$.03$	$+ .21$	$+ .101$
$.3364 =$	$.341$		

The fact that the quality of a scientist's research is far more important than quantity in receipt of awards is the first indication that the reward system in physics approaches its universalistic and rational ideal. However, research output explains less than half of the variance on awards. It is possible that there may be other variables, correlated with the number of awards received, which will indicate a gap between the real reward system and the ideal. Particularly important are variables characterizing the origins of scientists. A universalistic system will reward high-quality work without regard to the origins of the scientist producing it. The best available indicator of a scientist's scientific origins is the prestige of the place where he received his training. Some scientists in the

6. The results were the same when the average number of papers published per year, rather than total number of papers published, were used.

7. We were interested in seeing what results would be obtained by weighting the stratified sample so that it would be representative of the *top twenty departments*. There are not enough cases in the sample of 120 which can be weighted so as to make it representative of the entire population of university physicists. We therefore converted the sample back to a simulated random sample of the top twenty departments of physics by weighting each man according to the percentage of the population he represented in the stratified sample. Thus, a man drawn from a group representing 20 percent of the population was weighted twice as heavily as a man drawn from a group representing only 10 percent. This weighting procedure yielded results similar to those of the regression analysis. This suggests that the finding that quality is a far more frequent correlate of eminence than quantity may be generalized at least to the population of physicists in the twenty most prestigious departments.

sample of 120 received their Ph.D.'s from high-ranked prestigious academic departments and others had more humble origins. The zero-order correlation between prestige rank of a scientist's doctoral department and number of awards is .20. When quality is controlled, however, the standardized partial regression coefficient is .10. These data would lead us to conclude that reputational success is not influenced by a scientist's origins. Good work is recognized no matter where its producers come from.

Another variable of interest here is the rank of the scientist's current academic department. Of two physicists doing equal quality work, is the one located at Harvard more likely to receive recognition through awards than the one located at a less prestigious department? This is an important question to answer in assessing the extent to which the reward system is universalistic. However, there is a serious problem in answering this question. The way it has been posed assumes that rank of department is a potential determinant of receipt of awards. It is equally plausible, however, to believe that the number and type of awards a scientist has received (the amount of recognition) will have an effect on the rank of departments offering the scientist a job. In other words, the causal order between these two variables is indeterminate. The problem of unclear causal order comes up again and again throughout the analysis. The only way to handle the problem is to examine the interaction of the relevant variables over time. In the absence of such data we must be satisfied with the assumption that the various components of rank in the stratification system are related in a pattern of reciprocal causation. Thus, being at a high-ranked department may make a scientist more visible and increase the probability that he will be given an award. However, once a scientist has received an award he may be offered a job by a more prestigious department. We did have time data on a group of physicists between 1963 and 1966. Unfortunately, the stability of the two variables, number of awards and prestige of department, was so high in this short time-period as to make it impossible to use any of the several techniques available for establishing causal order.[8]

8. Rank of department at time 1 and rank of department at time 2 is correlated .82. Number of awards at the two times is correlated .94. For a discussion of how to use time data to get at causal order, see Donald C. Pelz and Frank M. Andrews, "Detecting Causal Priorities in Panel Study Data," *American Sociological Review* 26 (1961):854–66; O. D. Duncan, "Some Linear Models for Two Wave, Two Variable Panel Analysis," *Psychological Bulletin* 72 (1969):177–82.

RECOGNITION THROUGH APPOINTMENTS TO ACADEMIC DEPARTMENTS

The second kind of recognition for scientific work we shall consider is appointment to a major academic department, which we take as an indicator of positional success. Believers in the prevalence of the "publish or perish" policy hold that the number of publications determines appointments and that, even in the top-ranking universities, the scientist who has published only a few solid papers will typically be passed over in favor of the mass producer of trivia. Our data do not support this belief. The multiple correlation and regression analysis indicates that quantity of research output has an insignificant effect on the rank of a physicist's department affiliation, independent of the quality of his work. Once again, we have used a single-equation regression model to estimate the relative effects of the two measures of scientific output. Quality of research has a correlation of $r = .28$ with rank of department; quantity has a zero-order correlation of $r = .24$ with department rank. The multiple correlation between the dependent variable and quality and quantity is only $r/R = .29$. When we control for quality of work, there is only a minor independent effect of quantity on rank of department (partial regression coefficient = .11). Quality has a stronger net effect on rank of department ($b^* = .21$). In short, quantity of work has very little influence on the rank of a scientist's department.

The "publish or perish" belief is dealt a serious blow by the data presented. If we examine these data in tabular form, viewing the results as tentative because of problems inherent in the over-representation of eminent scientists in the sample, a further specification of our results becomes apparent (see table 1).

Type III physicists (publishing relatively few papers which are widely cited in the field) are the most likely to be in one of the top ten departments, even more so than the prolific Type I physicists.[9] But mass-producer Type II physicists fare no better than those who produce fewer, relatively undistinguished papers. The top departments, at least, prefer to choose their physicists on the basis of quality of research rather than on mere quantity.

9. A possible explanation of this finding is that those physicists who make a few important contributions, along with many run-of-the-mill contributions, may dilute their reputations. It may also be that, in the top departments, papers are more often circulated among colleagues, with the result that they are sometimes not published at all.

What is perhaps most striking about these data is not simply that quality once again proves to be a more significant determinant of social location in physics than quantity, but the apparent fact that neither of these variables actually accounts for a substantial amount of the variance on rank of department. Clearly, these variables have considerably less of a determinant effect on positional success than they do on reputational success. There are a number of reasons why the overall rank of a physics department might not be highly correlated with the quality of output of individual scientists. Positional success is more "voluntaristic" than reputational success. For many scientists, choosing an academic department involves criteria other than the department's prestige. It is possible that some of the

Table 1. Percentage of physicists in top ten departments of physics by quantity and quality of published research

	Quality	
Quantity	High	Low
High	Type I 58 (40)	Type II 29 (14)
Low	Type III 77 (22)	Type IV 27 (44)

Note: Quantity refers to the number of published papers. "High" quantity is 30 or more; "low" quantity is less than 30. High quality is 60 or more citations; low quality is less than 60.

physicists have chosen to affiliate themselves with departments with outstanding reputations for work in a particular specialty. Some physicists of the first rank simply choose to locate themselves at smaller universities or new university centers; others opt after some time for the quiet of the countryside over the pollution of the large cities, or a department may be chosen because a friend is working there or just because the scientist has been offered a higher salary. Such factors tend to produce a lower correlation between quality of work and rank of department than might be expected. There are probably some physicists in the sample who could get jobs at departments having higher prestige than their current ones. These findings do not imply that the top departments are not densely populated with physicists who produce outstanding work. They do suggest, however, that there are many talented physicists who are not employed at the most prestigious departments. High-quality research is being produced at many places in the social system.

Recently, Hargens and Hagstrom, and Crane have presented some data which suggest that an important factor in determining a scientist's positional success is where he started out, that is, the prestige of the department from which he received his doctorate.[10] After controlling for productivity, they found a significant effect of prestige of doctoral institution on current affiliation. Doctoral institution was as strong a determinant of current affiliation as productivity. These findings support earlier views presented by Caplow and McGee, who presented many qualitative examples of academic particularism.[11] Hargens and Hagstrom also found productivity to be a significant influence on stratification, but they offer us the possibility that institutional origins count as much as performance in determining a scientist's position in the scientific hierarchy. The findings of Hargens and Hagstrom seem to be inconsistent with the data presented above, where we found rank of doctoral department to be an insignificant determinant of reputational success (as measured by receipt of honorific awards). However, it turns out that our data, too, show that rank of doctoral department has an independent effect on rank of current department affiliation. The zero-order correlation is .29, and when quality is controlled the standardized partial regression coefficient is .25; the regression coefficient for quality of work is .24. Clearly, these two variables have approximately equal and independent effects on the rank of a scientist's current affiliation.

The data indicate that rank of doctoral department has a substantially greater influence on positional than reputational success. This is exactly what we would expect. Jobs are often obtained through leads provided by the professors with whom a scientist studied. Since professors at one high-prestige department are likely to turn to colleagues at other high-prestige departments either when they are seeking an additional faculty member or trying to place a former student, it is not surprising that rank of doctoral department has some independent effect on positional success. However, when

10. Lowell Hargens and Warren Hagstrom, "Sponsored and Contest Mobility"; Diana Crane, "Scientists at Universities."

11. Theodore Caplow and Reece J. McGee, *The Academic Marketplace* (New York: Basic Books, 1958).

scientists evaluate their colleagues, they generally consider their contributions to the field and not what school they studied at.

The multiple correlation of quality of work and rank of doctoral department on rank of current department is only r/R = .37, suggesting that other unspecified variables have an effect on the allocation of academic positions. Data that we briefly reported on in chapter 3 suggest that there is a significant independent effect of I.Q. on the prestige of a scientist's departmental affiliation after controlling for quality and quantity of research.[12] Perhaps skills associated with high mental aptitude, many of which become clear to colleagues through everyday interaction, are rewarded through appointment and retention at top departments. Further research is needed on this topic and on other variables which might influence a scientist's positional success.

RECOGNITION THROUGH VISIBILITY

Data bearing on a third form of recognition—familiarity of fellow scientists with one's research—were obtained by asking subsets of 1,308 physicists to indicate the extent of their first-hand acquaintance with the work of the prime sample of 120 physicists.[13] The extent to which each of the 120 was known to his colleagues is called his visibility score. A scientist's visibility score is simply the percentage of respondents who said they were familiar with his work.[14] This measure of visibility is dependent upon the subjective

12. See pp. 70–71.

13. See Appendix A. Since it was impractical to ask physicists to say how familiar they were with the work of the entire list of 120, five forms of the questionnaire, each form containing twenty-four names, were used. Having five forms of the questionnaire was equivalent to having the same study replicated five times. Although there were differences in magnitude of findings from form to form, in all cases the findings went in the same direction.

14. The exact question was: "These twenty-five physicists are drawn from various universities, institutes and fields of investigation. You may not know the work of some of these men, but please indicate, for each case, the extent of your familiarity with their work by circling the appropriate number: (1) familiar with most of his work, (2) familiar with some of his work, (3) familiar with a small part of his work, (4) unfamiliar with his work but have heard of him, (5) have never heard of him." We did not differentiate between the first three answers because we thought that scientists might use different frames of reference in estimating how much of a man's work they were familiar with. To check the adequacy of this decision, we computed mean visibility scores in which we did differentiate between all responses. The measure of visibility yielded correlation coefficients with the independent variables almost exactly like those presented in table 3.

Perhaps the most serious difficulty concerning our measure of visibility is that of determining the extent to which the physicists exaggerated their knowledge. As a crude gauge of the validity of response, a total of five fictional names was included

reports of the 1,308 physicists who returned our questionnaire. However, since we are more interested in relative differences in visibility than in the absolute scores, we believe that the exaggeration of some physicists' knowledge about the work of these men will not influence our conclusions.

Visibility of scientific research may be seen as a continuum ranging from one extreme occupied by the physicist who has produced no papers, and whose work is virtually invisible, to the other limiting case of the Wigners or Weisskopfs whose work is known to almost everyone in physics. Between these extremes lie the large majority of scientists who have published some research papers. The distribution of visibility scores is presented in table 2.[15] We now turn to a central question: why is the work of some physicists more visible than that of others?

Table 2. Visibility scores of the 120 physicists

Visibility scores	No. of men	%
90–100	13	(11)
80–89	8	(7)
70–79	7	(6)
60–69	8	(7)
50–59	9	(7)
40–49	11	(9)
30–39	17	(14)
20–29	12	(10)
10–19	20	(17)
0–9	15	(12)
	120	100%

Mean = 43.4
Standard Deviation = +29.3
Coefficient of Variation = 68%

We shall consider the probability that the work of different types of scientists will become known to the scientific community. For example, does the scientist who has consistently produced high-

in the list of physicists, one on each form of the questionnaire. The number of physicists reporting that they were familiar with the work of one of the fictitious men was 51, or 4 percent of the sample. Furthermore, of these 51, all but 3 were reporting on fictitious names which resemble closely the names of actual physicists.

15. Since our sample overrepresents eminent scientists, the average visibility score of university physicists generally can be assumed to be lower than the mean score of 43 for our sample.

quality work, but who is a member of a low-prestige department, have lower visibility than a colleague producing equal-quality work who is located at a major department? Or, correlatively, take the scientist located at an extremely prestigious institution whose work has not reached the standard set by his colleagues at that department. Does a halo shine over him by virtue of his location, giving his work greater renown than its quality might merit? What happens to the scientist who assumes he must publish or perish and so rushes into print at every available opportunity in search of recognition and, in passing, perhaps a prestigious academic position. Does this "mass producer" become highly visible? And finally, what of the scientists who produce only a few papers, all of which have been highly evaluated by the community of physicists?

Table 3 presents a matrix of zero-order correlation coefficients

Table 3. Intercorrelation matrix of visibility and possible determinants of visibility

		V	A	N	Q	R	P	D
V	Visibility	—	.67	.64	.58	.57	.49	.20
A	Prestige of highest award	—	—	.78	.40	.58	.34	.21
N	Number of awards	—	—	—	.57	.50	.45	.20
Q	Quality of output	—	—	—	—	.28	.60	.18
R	Rank of department	—	—	—	—	—	.24	.29
P	Quantity of output	—	—	—	—	—	—	.11
D	Rank of doctoral department							

N = 120

among visibility and several potential determinants of visibility. These several aspects of a physicist's scientific output and social location are highly related to visibility. The more awards a physicist has received and the greater their prestige, the more widely he is known. The more papers he has published and the more citations his work has received, the more widely he is known.[16] And finally,

16. The problem of causal order in this study is so difficult that even the time order of the relationship between visibility and citations is unclear. It is virtually certain that the publication of high-quality work precedes visibility in time. However, we are using citations as a measure of quality, and once a scientist has achieved visibility he will undoubtedly receive more citations. Thus the causal-order problem here is one of measurement. The measure of the independent variable, quality, is at least slightly contaminated by the dependent variable, visibility. If the difference

the higher the rank of the department at which he works, the higher his visibility. In short, those men who have produced the best research and who work at the best departments have higher visibility scores. We now want to see whether all these variables have independent effects on visibility.

Quantity and quality of output. One might expect that the sheer number of papers published by a scientist would influence his visibility. And, in fact, quantity of published articles is correlated with visibility. However, when we control for the quality of a physicist's work, the quantity of that work has only a slight independent effect on his visibility. The original correlation is r = .49; it is reduced to a partial regression coefficient of .22, when visibility is regressed on the quality and quantity of the scientist's work. On the other hand, quality has a strong independent effect on visibility after controlling for quantity of output ($b^*_{VQ.P} = .45$). The decomposition of variance on visibility, shown below, clearly indicates that the major portion of the explainable variance is a result either of the independent effect of quality of work or of the joint influence of quality and quantity of research on visibility. Once again, the sheer volume of scientific output has little influence on building a scientific reputation.

Decomposition of Explained Variance on Visibility

	Quantity	Quality	Quantity + Quality
$R^2_{V(QP)} =$	p_{VP}^2	$+ p_{VQ}^2$	$+ 2\, p_{VQ}\, p_{VP}\, r_{PQ}$
$(.61)^2 =$	$(.22)^2$	$+ (.45)^2$	$+ 2(.45)\,(.22)\,(.60)$
$.3721 =$	$.0484$	$+ .2025$	$+ .1188$
$.3721 =$	$.3697$		

If a physicist has produced high-quality papers, it makes little difference whether his bibliography is long or short. This indicates that the publication of scientific papers does not at all assure that their author will gain visibility. Papers not judged significant by fellow physicists, who therefore do not make use of them in their work, are functionally almost invisible.

Honorific awards. The publication of high-quality work increases the visibility of a scientist. We now want to examine the extent to which formal recognition of a man's work through the presentation

---

in the number of citations received by eminent and noneminent scientists was small, this would be a serious problem. However, since the differences are large, we can assume that the measurement problem has not had a substantial influence on the substantive conclusions.

of honorific awards increases his visibility. We have seen that there is a high correlation between a man's visibility and the prestige of the highest award he has received. The average visibility score of the eleven Nobel laureates in our sample was 85 percent. There is a considerable gap between this group and even such highly esteemed members of the physics community as the members of the National Academy, whose average visibility score was 72. A still sharper drop is seen when we look at the recipients of awards of lower prestige; their average score is 38. To complete the picture, those scientists without any honorific awards had an average score of 17.

Awards are given for high-quality work. Will a scientist who has produced such work, but who has not been formally recognized through the receipt of an honorific prize, be just as visible as his colleague who has been formally recognized? Regression analysis indicates that receiving formal recognition does add to one's visibility. Prestige of highest award has a strong, independent influence in explaining visibility, even after controlling for quality ($b^*_{VA.Q} = .52$). Here we face once again the problem of causal order. It is possible that visibility determines the granting of awards. To test this possibility we would need visibility scores at two points in time. Without such data we can only speculate that receipt of prestigious awards is likely to precede heightened visibility. Winning a prestigious award is likely to spread a scientist's reputation to the far corners of the social system.[17] Centrally located people are likely to know about important work before it is formally recognized through awards. In fact, to receive an award it is necessary that one's work come to the attention of the elite scientists who act as status judges.

Rank of department. So far the picture is reasonably clear. Physicists who do the best work and who receive formal recognition for it are most visible to their colleagues. Both quality of work and having honorific awards are variables characterizing the individual scientist. We now want to see how the physicist's visibility is influenced by the context in which he works. In a highly efficient communication system, work of high quality would be visible irrespective of where it was produced.[18]

17. In chapter 2, we presented data which showed that prospective laureates were more heavily cited than current laureates. Since winning the prize clearly increases visibility, these data indicated that producing high-quality work probably precedes high visibility although there is undoubtedly some feedback going on.

18. This employs the familiar technique of contextual analysis in which the relationship between the two variables characterizing individuals is examined in

The correlation between visibility and the rank of department of the physicist is .57. This alone, however, does not indicate that the stratification of departments impedes the smooth flow of information through the system.[19] The high correlation between visibility and rank of department may be due to a concentration of physicists who have produced high-quality work at high-prestige departments. To test this hypothesis, we must look at the relationship between rank of department and visibility, with quality of research output held constant. The data suggest that there is an independent and strong effect of both quality of work and the prestige-rank of department on the visibility of physicists.[20] When visibility is regressed on both quality of work and rank of department the partial regression coefficients are both statistically and substantively significant. The net effect of quality of output ($b^*_{VQ.R} = .46$) and of rank of department ($b^*_{VR.Q} = .45$) are roughly equal.

There are many possible explanations of the fact that being at a prestigious department enhances the visibility of scientific work. There is the interest of scientists on the periphery in keeping up with the developments at the center. Work being done at the major universities is therefore given more attention than its scientific significance might dictate. Most physicists, regardless of where they teach, received their Ph.D. degrees from one of the major departments. Even alumni who have moved to less prestigious departments are more likely to maintain an interest in what is being done at their alma mater. An additional explanation might be that the physicists doing high-quality work at top departments are more likely to work in highly visible specialties.[21] The data allow us to reject this interpretation. The relationship between rank of department and visibility remains, even when the variance due to specialty is removed.

---

different social contexts. See Paul F. Lazarsfeld and Wagner Thielens, Jr., *The Academic Mind* (Glencoe, Ill.: The Free Press, 1958), chapter 10.

19. It should be emphasized that we are dealing not with the communication of ideas but with a less specific type of information. One limitation of this study is that we cannot specify the precise character of information about the work of physicists.

20. Similar results were found by Crane, "Scientists at Universities," p. 710, in her study of biologists, political scientists and psychologists. Crane shows that rank of department and productivity have an independent influence on recognition. She sees this relationship as a result of the increased visibility of scientists at major universities.

21. See pp. 105–7 for a discussion of the visibility of physicists in different specialties.

Another explanation for the independent effect of rank of department on visibility would utilize the notion developed in chapter 3 of the subjective perception of quality. Perhaps the scientist located at a major department has wider visibility because his work is perceived as being of higher quality than that of a colleague doing "equal quality" work at a less prestigious institution.[22] Some data from another study allow us to make a tentative test of this hypothesis. The data are drawn from a study of 300 full professors in five disciplines at Ph.D.-granting institutions.[23] For each scientist we had a sample of his colleagues evaluate the significance of his work. The question was worded as follows: "Please indicate the relative importance of the work of the following scientists: has made very important contributions, has made above average contributions, has made average contributions, work has been relatively unimportant."

A scientist's mean score on this question can be used as a measure of his colleagues' subjective evaluation of the quality of his work. In this sample, rank of department was correlated with visibility $r = .43$. However, when we control for the quantity and quality of output and the subjectively assessed significance of work, the standardized partial regression coefficient for rank of department is .14. These data tentatively suggest that the effect of rank of department on visibility may be largely due to the tendency of scientists to evaluate the "same" work more highly if it is produced by a scientist located at a prestigious institution.

The results of our analysis of the determinants of visibility are summarized in table 4, in which we show the cumulative effect of quality, prestige of highest award, and rank of department on visibility. No other variable in the study increased the percentage of variance explained beyond this point. But perhaps this statistical procedure is masking some relevant substantive results.

Specialty. In science, as in most other institutions, certain activities have greater prestige than others. In physics, some specialties are more prestigious than others, and the subject matter of a man's work might influence his visibility. We compared the visibility of men working in the four largest specialties: atomic and molecular, elementary particles, nuclear, and solid state. Solid-state physicists had the lowest average visibility score, while physicists working in

22. Since we do not adopt an absolute definition of quality, by "equal quality" we mean work which is utilized to the same extent, as measured by citations.

23. For more information on this sample see Appendix A.

Table 4. Multiple regression of visibility on quality and quantity of scientific output, rank of department, and prestige of highest award

Sets of independent variables	Standardized regression coefficients	Zero-order correlation	r/R	R^2
1. Quality of output	.45	.58		
Quantity of output	.22	.49	.61	.37
2. Quality of output	.37			
Prestige of Highest Award	.52	.67	.75	.57
3. Quality of output	.46			
Rank of current department	.45	.57	.72	.52
4. Quality of output	.36			
Rank of department	.25			
Prestige of highest award	.38		.78	.61

N = 120.
NOTE: All of the standardized partial regression coefficients are statistically significant at greater than the .01 level, except the partial regression coefficient for quantity of output, which is significant at the .05 level.

elementary particles had the highest average score.[24] This relationship was maintained even when quality of work was taken into consideration (see table 5). Men working in particle and nuclear physics still have the highest visibility.[25] Atomic and molecular physicists have somewhat less visibility and solid-state physicists have the least visibility. Physicists who have produced "low quality" work in elementary particles are just as visible as solid-state physicists who have produced "high quality" work. These data indicate that reputation for work done in so-called "hot fields" can more easily transcend the boundaries of specialty than work in less prestigious fields. Does the fact that specialty has a moderate independent effect on visibility indicate a breakdown in universalism? If physicists believe that the problems which members of one specialty address are more important than those addressed by members of another specialty, then it would not be a departure from rationality. Im-

24. The reader will note that we have switched here to tabular analysis. A dummy variable regression analysis of these data would have been possible, and surely statistically as adequate, but we felt that the reader could more easily interpret these tabular results than the regression results when these nominal scales such as specialty were part of the analysis.

25. Mean visibility score was computed by counting the number of positive identifications of scientists in the group and dividing by the total number of possible identifications.

pressionistic evidence suggests that in recent years physicists have seen the most important problems in their discipline as those in elementary particles.

Table 5. Mean visibility score by specialty and quality of work *

Specialty	Mean Visibility Scores, Quality: High (60+ citations)	Mean Visibility Scores, Quality: Low (0–59 citations)
Elementary particles	64	40
Nuclear	61	32
Atomic and molecular	50	16
Solid state	39	14

	Quality: High		Quality: Low	
	No. of men	Total no. of identifications	No. of men	Total no. of identifications
Elementary particles	10	2,658	11	2,861
Nuclear	12	3,163	26	4,357
Atomic and molecular	13	3,418	1	272
Solid state	6	1,584	6	1,547

* The number of physicists in each group and the total number of possible identifications of men in each group are presented below. The mean visibility score for any group can be seen as the percentage of positive identifications.

Age. The correlation between age and visibility is .03. We thought that this negligible correlation might be the result of a curvilinear relationship between the two variables. This turns out to be the case. Physicists under the age of forty have a mean visibility score of 36. Visibility seems to increase with age and reaches a peak score of 54 in the early sixties. However, those physicists who were sixty-five years or older had an average score of 37. These data indicate that there tends to be a substantial time lag in the growth of scientific reputation. A young man may produce significant work, but his work might not become highly visible for several years. Table 6 shows that this is not the case. Among those scientists who have produced high-quality work, age has no effect on visibility.[26] Younger men

26. We thought that this finding might be due to the dichotomy on quality. Perhaps young physicists of the first rank become immediately visible and the good (but not brilliant) young physicists gain visibility with age. However, when we split

who have published high-quality work are just as visible as their older colleagues who have produced high-quality work.[27] The fact that young men who have contributed to their field are just as visible as their older colleagues who have made contributions of comparable quality is another indication of the highly effective communication system in physics. Table 6 also indicates that even the scientist nearing or actually in retirement retains his visibility if he has produced high-quality work. On the other hand, the scientist who has not produced such significant work loses visibility rapidly as he approaches retirement. These scientists, who had some visibility due to the sheer volume of their activity in the field, fade into anonymity once they are no longer active, and younger scientists no longer experience them as significant parts of their environment.

Since our data measure visibility at only one point in time, we could not directly study how different career patterns influence visibility. However, we were able to divide those scientists who have produced high-quality work at some point in their career into those whose work is still significant and those whose work is not.[28]

---

our high-quality group into those with more than 100 citations and those with 60 to 99 citations, the results did not change. Quality, not age, determines visibility. A more striking demonstration of this is obtained by using prestige of highest award as the control variable. The few young physicists who had received top honors had a considerably higher average visibility score than their older peers. The data are presented below.

Mean visibility score

	Prestige of Highest Award		
Age	High	Medium	Has No Awards
65 or older	72	24	12
60–64	74	46	37
50–59	77	32	14
40–49	77	46	21
Under 40	94	36	39

27. The curvilinear relationship between age and visibility may result from the correlation between age and quality in this sample (r = .14). This correlation is totally a result of the sampling criteria used to choose the 120 physicists. Since we were interested in including many scientists who possessed honorific awards, we included many elder eminent physicists. Among the 1,308 physicists who returned our questionnaire, there was a slight negative correlation between age and quality.

28. To determine whether a physicist had ever produced "high quality" work, we used the technique of weighting citations described in chapter 2. Sixty-two of the 120 physicists had 60 or more weighted citations in their three best years. We then classified these men by the total number of unweighted citations (a measure of the current significance of their work) in the years 1961, 1964, 1965. Forty-nine of the 62 physicists had 80 or more unweighted citations.

The first group had an average visibility score of 64, and the latter an average score of 46. This suggests that just as a man may become visible at an early age if he has made a substantial discovery, his visibility declines if his early work loses significance and his current work does not meet the standards set by his earlier performance.

Table 6. Mean visibility score by age and quality of work *

Age	Mean Visibility Score — Quality: High (60+ citations)	Mean Visibility Score — Quality: Low (0–59 citations)
65 or older	60	13
60–64	62	35
50–59	60	29
40–49	59	31
Under 40	58	24

Age	High: No. of men	High: Total no. of identifications	Low: No. of men	Low: Total no. of identifications
65 or older	11	2,893	11	2,864
60–64	10	2,662	6	1,543
50–59	21	5,444	13	3,387
40–49	12	3,190	12	3,126
Under 40	8	2,073	16	4,210

* The number of physicists in each group and the total number of possible identifications of men in each group are presented below.

Collaboration. Another variable possibly influencing visibility is the position of an author's name in a series of collaborating authors. Today, collaborative research is the modal type in physics. It is possible that a physicist who published few noncollaborative papers or whose name rarely appeared first among collaborators might be less visible to his colleagues. Recent research by Zuckerman has shown that Nobel laureates are apt to claim that name-ordering of authors is an insignificant matter. However, they often add that name-ordering is a source of stress and conflict among coworkers. This problem is often circumvented in physics through the alphabetical arrangement of names.[29] Nonetheless, some scientists feel

29. About 60 percent of the multi-author papers listed in the *Abstracts* since 1920 have alphabetized name orders. See Zuckerman "Patterns of Name Ordering."

that being first or last in a series of names will make their work more visible. Here we focus on only one question related to name-ordering and recognition: in the end, does a *C*ain's visibility to his fellow physicists suffer at the hands of an *A*bel? This question is more complex than it at first appears; consideration should be given the varying eminence of the collaborating authors and to various alternative forms of name-ordering. Nevertheless, tentative evidence leads us to believe that name-ordering is of more concern to the authors of papers than to the people who read them, and has little effect upon the ultimate visibility of a scientist. The proportion of a physicist's work on which he is either the first author or the only author has a correlation of .17 with visibility; however, when quality is controlled, the standardized partial regression coefficient is .04. We may conclude that regardless of how often a scientist's name appears alone or first among a group of collaborators, his visibility will be high only if the quality of his research is high.[30] The position of an author's name on a paper has no independent effect on his visibility when the quality of his research is known. Let us stress that these data deal only with the visibility of a scientist's entire body of work and not with any specific piece. Although name-ordering may have negligible effect on a scientist's ultimate visibility, it may be important in assigning credit for a particular piece of work.[31]

Rank of doctoral department. Once again, we must consider the extent to which a form of recognition, visibility, is influenced by the scientific origins of the scientist. If the previous analysis is correct, we would expect rank of doctoral department to have little influence on visibility. The zero-order correlation is .20; however, when quality of work is controlled, the standardized regression coefficient is reduced to .10. This suggests that rank of doctoral department has virtually no independent effect on visibility.

REINFORCEMENT OF RESEARCH ACTIVITY BY THE REWARD SYSTEM

One of the major findings of this research has been that the quality of a scientist's research is more important than sheer quantity in

30. We are aware of the difficulty involved in including single-author papers with collaborative papers on which the author's name appears first. Because it is not possible to separate the amount of visibility due to each of these types of papers, it is necessary to consider them jointly.

31. The distinction between the visibility of a physicist's entire body of work and the visibility of any particular piece of work is the same as the distinction between the communication of specific ideas and the communication of more general knowledge of a man's work. In this analysis we deal only with the latter type of communication.

determining location in the stratification system. Since quality and quantity of research output are fairly highly correlated, the high producers *tend* to publish the more consequential research. There are at least two basic processes creating the high correlation between quantity and quality of work. The gist of the matter is that engaging in a lot of research is in one sense a "necessary" condition for the production of high-quality work. As scientists of the first rank remind us, producing significant science is a risky business, full of uncertainties.[32] There is seldom a guarantee that a program of research will produce important results, and do so in short order. A physicist will try an idea out, and sometimes it will work but more often it will not. It is the rare scientist whose eye for the crucial problems is so keen that he limits his energies solely to an investigation of these problems. Even the "average" top scientist must make many experiments before he gets an exciting result. We believe that unless a physicist makes a large number of attempts (that is, has high productivity) the probability of his making a significant discovery will be low.

A second and more important reason why quality and quantity of work are so highly correlated is that the reward system operates in such a way as to encourage the creative scientists to be productive and to divert the energies of the less creative scientists into other channels. We shall here analyze the processes through which the reward system works to create and then reinforce the correlation between quantity and quality of work.

Most physicists have been trained in the major departments. Fifty-six percent of the physicists among the 1,308 we queried received their doctorates from the top-ranked fifteen departments; 44 percent from the top ten. We assume that, as students, they internalized the prevailing norm of doing research in these departments. Soon after they receive their degrees (and sometimes before), these young physicists begin to publish their research, whether alone or as part of a research team. Their papers must first pass through the evaluation system. The first screening of this work is by the referees associated with the journals; the prime journal in the field, the *Physical Review,* for example, has an especially elaborate system of refereeing papers. Standards are high, and the manuscripts even of eminent scientists are sometimes rejected.[33] Even though

32. See Harriet A. Zuckerman, "Nobel Laureates in the United States" (Ph.D. dissertation, Columbia University, 1965), for the reports of laureates on the "pedestrian research" that often precedes scientific results of significance. See also Michael Polanyi, *Personal Knowledge* (London: Routledge & Kegan, Paul, 1958).

33. Zuckerman and Merton, "Patterns of Evaluation."

standards employed by the journals are high, a majority of papers are accepted for publication. More important than the formal evaluation of journals is the informal evaluation after publication by the international community of physicists. Sometimes the published paper is largely ignored, with few citations to it, or it may be identified as a significant contribution and put to use in many other published researches. If the reward system, in the form of recognition by citation, does affect research productivity, we assume that the greater such collegial recognition of these early researches by physicists, the greater the probability that the physicists will continue to be productive. We hypothesize that few scientists will continue to engage in research if they are not rewarded for it.[34]

To test this hypothesis, we have traced the sequence of publication patterns among the sample of 120 physicists. We divided them into two broad categories: the early producers, who published three or more papers within five years after the Ph.D.; and the others, who published fewer than three. We then examined the collective responses to these early publications as measured by the number of citations they received within the same five-year span.[35] Finally, we compared the later productivity of the physicists receiving differing amounts of recognition in these early years. The results are presented in table 7. Three-quarters of these physicists began their professional careers by publishing at least three papers soon after their doctorates. There are few "late bloomers": only five of the thirty physicists who started off slowly ever became highly productive (averaging 1.5 or more papers a year).

Consider the sequence of publication patterns among physicists

34. Storer suggests the importance of recognition in reinforcing motivation for scientific research: "recognition is frequently interpreted by him [the scientist] . . . not only as confirming the validity and significance of his work but more generally as an affirmation of his own personal worth, and it thereby gains meaning as an intrinsically satisfying reward. Further . . . the act of creativity does not seem to be complete without some feedback from others, and in science, this feedback takes the form of recognition. Recognition is thus the appropriate response to creativity and is of *significant importance in the desire to engage in research*" (emphasis added) (Norman W. Storer, "Institutional Norms and Personal Motives in Science" [Paper presented at the annual meeting of the Eastern Sociological Society, April 1963]).

35. Citations are being used in a slightly different sense here. From the standpoint of the system of science, citations indicate the impact of a piece of research; from the standpoint of the individual scientist, citations to his work provide a type of recognition. Since most of the 120 physicists received their degrees quite a number of years before the citation index was established, we have used the technique of weighting 1961 citations as an indicator of how many citations early work received (see chap. 2).

who started their research careers by being productive. The more citations their early work received, the more likely they were to continue being productive. Only 30 percent of those who received 0–25 weighted citations went on to continued high productivity, in contrast to the 76 percent with more than 100 weighted citations. These findings suggest that when a scientist's work is used by his colleagues he is encouraged to continue doing research, and that when a scientist's work is ignored his productivity will tail off.[36]

One obvious problem with the data presented in table 7 is that the measure of early citations (recognition) is retroactive. It is possible that those men who continued to be productive received more citations in 1961 to their early work than they did at the time the work

Table 7. Percentage of physicists who continue to be productive, by early recognition and early productivity

Early recognition (number of weighted citations received in first five years after Ph.D.)	Early productivity	
	Wrote 3 or more papers in five years after Ph.D.	Wrote less than 3 papers in five years after Ph.D.
0–25	30 (27)	15 (27)
26–100	48 (38)	33 (3)
101 or more	76 (25)	.. ..

NOTE: Base of percentage is in parentheses. Since the percentages are based upon relatively small numbers, we computed chi-square for the first column. $X^2 = 11.35$, d.f. 2, $P > .01$. Continued productivity was defined as producing 1.5 or more papers per year after the immediate postdoctoral years.

was published. To make a methodologically sound test of the hypothesis that the high correlation between quantity and quality was produced by the operation of the reward system, it was necessary to collect additional data. We took a random sample of fifty physi-

36. Table 7 shows how the correlation between quantity and quality is reinforced by the least institutionalized part of the reward system—the use of one's work by colleagues. We believe that the more institutionalized parts of the reward system work in the same way. Thus, we would hypothesize that those scientists who have received some award in the early part of their careers are more likely to continue being productive than their colleagues who have not received awards. Also, those young scientists whose doctoral and immediate postdoctoral work is recognized by appointment as an assistant professor at a top university are more likely to continue producing than their colleagues who become assistant professors at the lesser departments.

cists who received their Ph.D.'s from American universities in 1957 and 1958.[37] We then looked up in *Science Abstracts* the number of papers each had published in every year from the time they received their degree through 1969. We were thus able to compare productivity in the first five years after the Ph.D. with productivity in the last five-year period (1965–69). We were able to use citations received in 1961 as an indicator of recognition received for early work.[38] The data presented in table 8 fully confirm the interpretation given to table 7.[39] The later productivity of physicists was substantially influenced by recognition received by their early work. Those men who produced the best work were rewarded and continued to be highly productive. The productivity of those who were not rewarded dropped off.

Table 8. Mean number of papers published in 1965–69, by early recognition and early productivity*

	Early productivity	
Early recognition (number of citations received in 1961)	Wrote 3 or more papers in five years after Ph.D.	Wrote less than 3 papers in five years after Ph.D.
0	3.0 (6)	1.3 (12)
1–10	4.0 (16)	2.6 (7)
11 or more	9.6 (8)	† (1)

*The results were exactly the same where we used the proportion producing 3 or more papers in 1965–69 as the measure of the dependent variable.
†Too few cases to compute.

Our hypothesis that a reward system which rewards quality of output produces a high correlation between quantity and quality might be called the reinforcement, or behaviorist, theory. Simplified, it states that scientists who are rewarded are productive, and scientists who are not rewarded become less productive. An alternative explanation of the data, one held by many scientists and historians of science, might be called the "sacred spark" theory.

37. See *Dissertation Abstracts.*
38. We looked up citations to all papers published by each physicist, regardless of whether he was first author.
39. The statistics in table 8 are based upon a small number of cases. We have replicated this table on the sample of 499 men and women scientists (see Appendix A). The results were almost exactly the same.

Adherents of this theory would argue that scientists do science not because they are rewarded but because they have an inner compulsion to do so. Certainly the history of science offers many examples of men like Cavendish who shunned external rewards but continued throughout their lives to do a great deal of excellent science. According to the sacred-spark theory scientists who are rewarded and continue to be productive would be productive even if they were not rewarded. We would argue that both theories might be correct. Some scientists would undoubtedly continue to be productive even if they were not rewarded, and others would not. However, to split apart the amount of variance explained by each of the theories would be difficult if not impossible. For once a scientist has been rewarded, how can we say whether he would have continued to be productive if he had not been rewarded?

If recognition motivates scientists who produce high-quality work to be productive, why is it that we find some physicists (Type II-mass producers) who produce large numbers of relatively trivial papers? It is possible that some of the mass producers could have the sacred spark. We would like to suggest, however, that criteria of evaluation are not applied equally through every sector of academe. It is likely that there are some departments in which people will be rewarded for the sheer quantity of their output. We would expect this pattern of reward to occur more frequently in the relatively low-prestige departments. If this were true, we would find a higher correlation between the quantity and quality of research output in high-prestige departments than in those of lesser prestige. The data support this hypothesis. For those of the 120 physicists working at distinguished departments the correlation between quantity and quality was $r = .71$; for those working in other departments the correlation was $r = .42$.

These correlations may result from a process of social selection whereby the outstanding departments recruit abler researchers with a better sense for the significant research problem. On the average, then, their research has a greater impact on the field. The weaker departments, further removed from the springs of scientific advance, tend to recruit less able investigators who gradually lose contact with the rapidly advancing frontiers of physics and produce work of less significance. Some of the physicists in these departments continue to publish, and it is in these departments, which find it difficult to recruit faculty who do research at all, that the sheer number of publications is more apt to be an important criterion for promotion. Thus, the reward system of the weaker departments more often makes for that displacement of goals which is expressed in the policy of "publish or perish." The findings might also be interpreted to result from an imperfect communication network in science. This interpretation assumes that the flow of scientific information moves principally in a one-directional path from the major to the minor centers of scientific inquiry. Physicists in the weaker departments are apt to know more of the work of men in the stronger departments, while men in the stronger departments less often monitor the work of physicists in the weaker departments. Work of "equal quality" produced in departments of differing rank will be differentially recognized and cited in the field. An extension of this view of the communication network maintains that the leading journals of physics are "controlled" by the same group of men who control the top-rated departments. The journals more readily publish papers by members of the in-group and their students, who tend to cite the work of others in the group. This results in differing citation rates for research of comparable significance published by physicists located in departments of differing prestige. We present data bearing on these interpretations in the following chapters.

THE SCIENTIFIC REWARD SYSTEM

Thus far we have been analyzing various aspects of the reward system separately. We now want to consider several aspects together. Because a multiplicity of variables interact to determine a scientist's location in the stratification system, it is difficult to characterize the system without considering the simultaneous effect of all the significant variables. What makes this difficult is the fact that in science variables influencing the process of stratification interact with each other. Because of a multiplicity of patterns of reciprocal causation, it is difficult to estimate path coefficients for a realistic model depicting the operation of the system. In figure 2 we present the model which we believe most accurately represents the operation of the scientific reward system. We cannot estimate the path coefficients for this model because we do not have information on all the variables for any one sample and because of the three reciprocal paths in the model. We shall therefore discuss this model as a theoretical one.

At the far left in the model are the prestige of the department at which the scientist earned the doctorate, and I.Q. as a measure of native ability. In a completely universalistic system, rank of doctoral department or social origins would have no independent influence on the distribution of rewards. Physics, of course, is not completely universalistic, and rank of doctoral department does have a moderate independent influence on the rank of the physicist's current academic affiliation, which is taken here as an indicator of positional success. As the absence of a path from rank of doctoral department to visibility indicates, the social origins of a physicist has no independent effect on reputational success. There are several possible explanations of the independent effect of social origins on positional success. It could be a result of the informal sponsor system that exists in academe. The sponsor system works most strongly in the attainment of a first academic position. A graduate

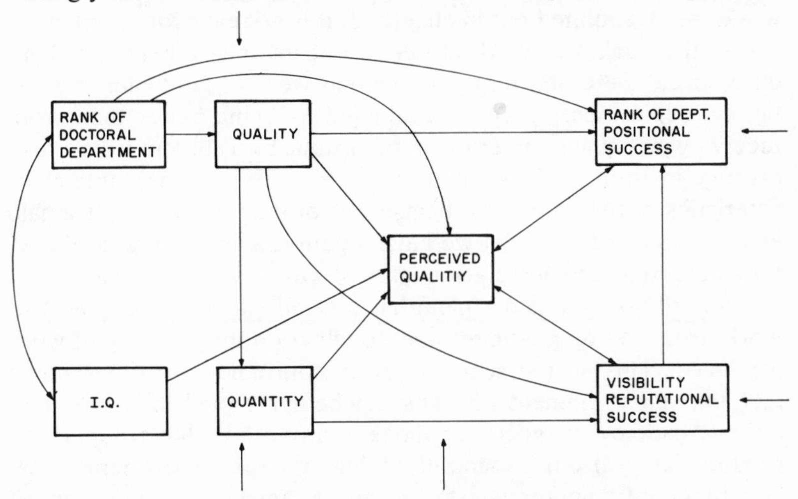

FIG. 2 THEORETICAL MODEL FOR SCIENTIFIC REWARD SYSTEM

student's sponsor will make contacts leading to the student's employment. Since sponsors at top-ranked departments are more likely to have contacts at other top-ranked departments, their students have an easier time getting jobs at prestigious institutions. Analysis of other data we have collected indicates that the effect of rank of doctoral department on prestige of current department is almost entirely mediated by the influence of rank of doctoral department on a scientist's first job. Of course, the prestige of a scientist's first job has a strong independent effect on the prestige of current job. We may conclude that sponsorship is primarily important in getting a first job and has little effect on later job acquisition. Also, once a scientist gets into a specific rank in the stratification system, structural processes operate to keep him there. These processes are exemplified by the different criteria usually employed for promoting insiders and hiring outsiders. Because insiders are given points for service to the department and university, and for teaching, that are usually not given to outsiders, the former can be promoted on the basis of a considerably weaker publication record than that expected of outsiders. Likewise, it is difficult for good scientists below the top level to make the necessary contacts to be upwardly mobile from low-ranked to high-ranked departments.

There is at least one other compatible explanation of the effect of rank of doctoral department on current department. This explanation involves the subjective perception of the quality of a scientist's work. As we pointed out in chapter 3, it is not easy for scientists to assess the quality of work of two scientists unless there is a sharp difference. Therefore, it is possible that work of graduates of prestigious departments may be evaluated as being better than work receiving an equal number of citations done by a student from a low-prestige institution. This could result from either unmeasured characteristics of the graduates themselves, or it could be merely a halo effect. Thus, in the model we have a path drawn from rank of doctoral department to perceived quality of work.

Rank of doctoral department has a small effect on the quality of work produced by graduates and no effect on the quantity of work produced. This at first seems to be in contradiction to the known fact that most eminent physicists, whether Nobel laureates, National Academy members, or those in any other elite group, come predominantly from a handful of high-ranked departments. The apparent contradiction vanishes when we remember that top-ranked departments produce the majority of Ph.D.'s. Thus, although it is true that most eminent scientists come from major departments, it is also true that most noneminent scientists come from these same departments.

I.Q., used as an indicator of native ability, is correlated with the rank of doctoral department. The graduates of top-ranked departments have higher I.Q.'s than graduates of low-prestige institutions. However, there is no relationship between I.Q. and either the quality or quantity of work produced. It might seem then that I.Q. would not be a significant variable in the scientific reward system. But I.Q. is correlated with the rank of a scientist's current academic department, independently of prestige of doctoral department. We would hypothesize that the relationship between I.Q. and rank of current department would probably be mediated by perceived quality of work. Scientists with high I.Q.'s are probably better able to project positive self-images and therefore have their work perceived in a favorable light.

In figure 2 we have drawn a path from quality to quantity of work, indicating a causal connection. As our analysis of the relationship between these two variables indicated, producing high-quality work reinforces research activity. Those who have early success are rewarded and motivated to do more research. In this way the reward system creates the high correlation between the quantity and quality of work produced. One result of the empirical analysis we have conducted that we can be unequivocal about is that in physics sheer quantity of publication has virtually no independent effect on positional success and only a small independent effect on reputational success. Producing a large number of trivial papers has little utility in gaining recognition in physics. Quality of work, on the other hand, does play a major role in the reward system.

The effect of quality on rewards is mediated by the perceived quality of work. The independent effect of quality on positional and reputational success is slight. This allows us to make a subtle but important modification in analyzing the way in which universalism works in science. There can be little doubt that the quality of work *as it is perceived by other scientists* is the most important variable in determining the allocation of rewards. There is also little doubt that to have one's work highly evaluated you must actually produce work that other scientists find useful, that is, work which is highly cited. The production of work deemed useful is a necessary condition for earning a reputation as a good scientist. It is also true, however, that variables other than the actual substantive content of the work produced, influence the subjective evaluation of one's work. Two of these variables are the rank of doctoral department and intelligence as it becomes evident in interaction. Rewards the scientist has received are other variables influencing the subjective perception of quality. This brings us to a consideration of the process of accumulative advantage.

In almost every case in which the scientific reward system seems to depart from universalism, we can find the operation of accumulative advantage. In science, as in other areas of life, those who are initially successful have greater opportunities for future success. Students who get into good graduate departments have better chances of getting good jobs. Scientists who are perceived as having done high-quality work receive both positional and reputational success. But scientists who are successful are more likely to have their work perceived favorably, independently of the content of the work. As the reciprocal paths in the model indicate, there is also feedback between the two forms of recognition themselves. Scientists who are at top-ranked departments will be more visible, independently of other variables, and scientists who are visible are more likely to get jobs at top departments, independently of other variables. We may conclude that doing good science brings rewards, and once those rewards are received they have an independent effect on the acquisition of further rewards. It makes little difference whether positional or reputational success comes first, the attainment of one aids in the attainment of the other.

Although we cannot estimate the path coefficients for the theoretical model of how the scientific reward system works, we have two different sets of data which allow us to look empirically at two approximations of this model. In figure 3 the model is based on data

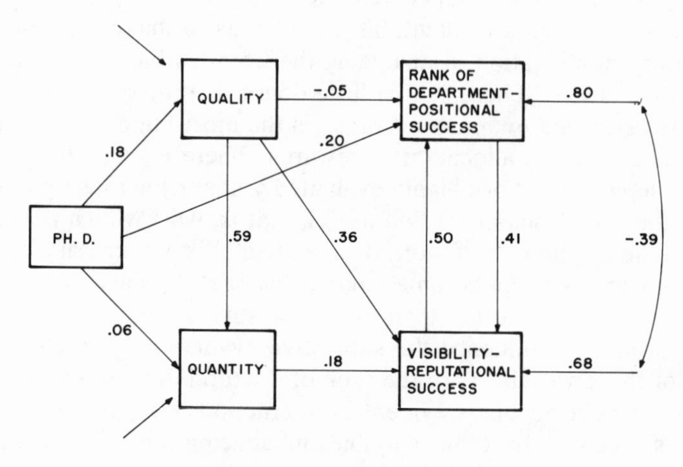

FIG. 3 THE SCIENTIFIC REWARD SYSTEM
(120 PHYSICISTS)

characterizing the sample of 120 physicists. The path coefficients lend support to the discussion of the theoretical model. Particularly worth noting is the fact that among the 120 physicists quality of work, as measured by citations, has no independent effect on rank of department. In fact, the path coefficient is negative. All the influence of quality of work is mediated by visibility. Producing high-quality work will make a scientist visible, and visibility will lead to appointment at prestigious departments. Also of interest are the size of the two paths connecting positional and reputational success. Here we have tentative empirical evidence to support our accumulative advantage argument. Independently of the other variables in the model, positional and reputational success influence each other.

An even better approximation to the theoretical model can be made by using data drawn from the sample of 300 full professors at Ph.D.-granting institutions in five fields. For this sample, as we noted above, we have a measure of the perceived quality of work. The only ways in which the model presented in figure 4 differs from

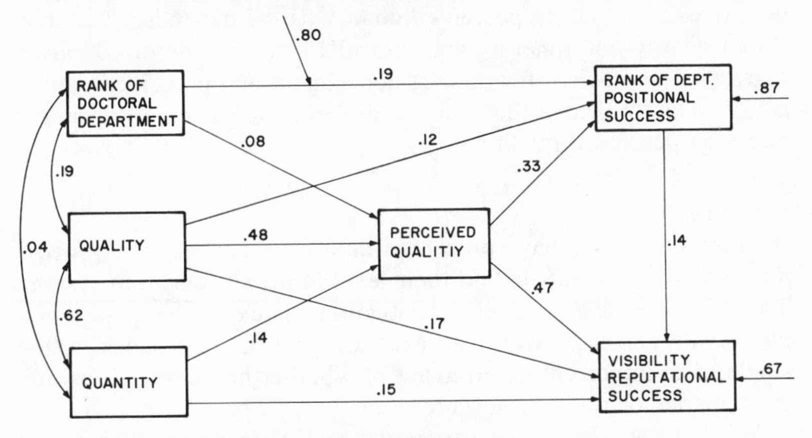

FIG. 4 THE SCIENTIFIC REWARD SYSTEM
(300 FULL PROFESSORS AT PH.D. GRANTING INSTITUTIONS IN FIVE FIELDS)

the theoretical model of figure 2 is that we have no data on I.Q., we have allowed quality and quantity to be intercorrelated, and we have replaced reciprocal paths with unidirectional paths.[40] Figure 4 offers strong support for the interpretation given figure 2. Quality of

40. We tried to estimate the paths for several models, including reciprocal paths. In each case the model yielded an "unbelievable" path coefficient. For example, when we had reciprocal paths connect positional and reputational success, the path from positional to reputational success was .49, and the path in the opposite direction was −.23. Given that there is no logical explanation for such a result, although there is a statistical one, we chose to use a model with unidirectional paths as an approximation of the theoretical model of figure 2.

work, as measured by citations, is the key determinant of perceived quality. Rank of doctoral department and quantity have only slight effects on perceived quality. Most of the influence of quality on positional and reputational success is mediated by perceived quality, which in turn has a strong influence on the two forms of success.

Particularly interesting is the relatively weak independent influence of positional success on reputational success. This suggests that perceived quality is the mediating variable in the operation of accumulative advantage. Positional success does not directly bring reputational success. It does bring an increase in the perceived quality of work, which in turn will bring an increase in reputational success. Although we would, of course, need data over time to actually establish the validity of this argument, we can make a tentative test by using perceived quality as the dependent variable. When quality and quantity are controlled, the standardized partial regression coefficient for rank of department and perceived quality is .27. This indicates that rank of department has an independent effect on perceived quality.

CONCLUSION

In this chapter we have presented data on aspects of the process through which scientists find their level in the stratification system. The main conclusion of our investigations is clear: the quality of a physicist's scientific work, as evaluated by his colleagues, is the single most important determinant of whether he rises to a position of eminence or remains obscure.

We have studied three separate types of recognition accorded scientists: receipt of prestigious awards, appointment to prestigious departments, and having one's work widely known by colleagues. In all cases, quality of research was a key variable. In fact, for the two nonvoluntaristic forms of recognition—receipt of awards and visibility—quality of work explained a substantial amount of the variance.

The reward system of science, of course, is not completely universalistic. We have shown that the prestige of the department from which the Ph.D. was obtained does have an independent effect on the three forms of recognition, just as rank of a man's current department has an independent effect on his visibility. Despite these deviations from the ideal type, it would probably be safe to conclude that in no other major social institution is the process of stratification more universalistic.

5 DISCRIMINATION AGAINST WOMEN AND MINORITIES IN AMERICAN SCIENCE

In the previous chapter we analyzed how a scientist's rank in the stratification system was influenced by a set of variables characterizing his position within the institution of science. We noted, for example, how the prestige of the department in which the scientist was trained influenced his subsequent career. It is, of course, possible that scientists could be discriminated against on the basis of their nonscientific statuses. In a truly rational and universalistic system, statuses such as race, sex, and religion, which are by assumption functionally irrelevant for the performance of scientific roles, would have no independent influence on the distribution of rewards. There is widespread belief and much vociferous protest that science does not approach this ideal. Critics of science jump from the known fact that the proportion of women and black scientists is much lower than the proportion these groups represent in the population at large to the assumption that discrimination is occurring. In this chapter we analyze the extent to which such discrimination does, in fact, occur.[1]

To assess the presence of discrimination is a far more complicated problem than at first appears. Obviously, not every person in the population is qualified to be a scientist. Therefore, comparing the proportions of a particular group in the population and in science tells us little about the extent of discrimination. If we assume that possession of a Ph.D. is a prerequisite for an academic science job, then we must examine the distribution of rewards only among those

1. There are two analytically distinct dimensions to a study of discrimination: behavioral and attitudinal. Prejudicial attitudes are not necessarily accompanied by discriminatory behavior. Robert K. Merton discusses the typology that emerges from these two aspects of discrimination in "Discrimination and the American Creed," in R. M. MacIver, ed. *Discrimination and National Welfare* (New York: Harper and Brothers, 1949), pp. 99–126. In this chapter, we do not examine the attitudes of scientists toward women or minorities within their ranks. Here we look only at the consequences of these attitudes for recognition accorded to women, blacks, and Jewish scientists.

with Ph.D.'s. In other words, the fact that a particular group is underrepresented in science could be the result of social processes at work outside of science. Two types of processes must be considered: social selection and self-selection. An example of how these two processes could influence the proportion of any group entering science might be found in graduate school admissions. If graduate schools were systematically rejecting qualified women and blacks, this would certainly constitute discrimination—exclusion of these groups from science by social selection. However, small proportions of women and blacks in graduate schools may also be a result of self-selection. If women and blacks choose not to apply for admission to graduate school, we cannot jump from the fact of their underrepresentation to the conclusion of discrimination. The failure of women and blacks to apply to graduate school may be explained by discrimination occurring earlier in life or by a value system influencing the career choices made by members of these groups.

Although it is obvious that there is a good deal of discrimination against women and racial minorities in the United States, this does not mean that such discrimination occurs in every institution.[2] The question which we begin to examine here is the extent to which discrimination on the basis of functionally irrelevant statuses occurs in science *after the receipt of the Ph.D.*[3] Even if we were to find no evidence of discrimination, it would, of course, be possible that discrimination was occurring in science prior to the receipt of the Ph.D. One task of the sociologist is to locate the exact points where discrimination does or does not occur. We shall present data on the extent to which women, blacks, and Jews are discriminated against in university science departments. A thorough examination

2. There is, of course, a vast literature on the problem of discrimination in American society. See, among many others, Gunnar Myrdal, *An American Dilemma* (New York: Harper Torchbook, 1944), and H. M. Blalock, Jr., *Toward a Theory of Minority Group Relations* (New York: John Wiley and Son, 1967).

3. Functionally irrelevant statuses have been discussed at length in the lectures of Robert K. Merton at Columbia University. They refer to those status characteristics of individuals which are unrelated to the ability to perform adequately specific roles. It is not always easy, of course, to say with certainty whether a status characteristic is or is not functionally relevant. If a professional basketball team uses sex status as a criterion for the selection of individuals for the team, is a functionally irrelevant status being imported into the situation, or is sex status relevant to adequate performance? For one study of women which has used this concept, see Cynthia F. Epstein, *Woman's Place: Options and Limits in Professional Careers* (Berkeley: University of California Press, 1970).

of this problem would require far more and better data than is currently available. Here we present what are essentially preliminary results of an ongoing research project. Most of our data are on women in science, and this is the first topic we shall consider.

SEX STATUS AND SCIENTIFIC RECOGNITION

Throughout its long history, science has been populated almost exclusively by men; until the twentieth century, the phrase "men of science" could be taken as almost equivalent to the tag "scientists." Even today the situation has not drastically changed: there is a striking paucity of women in science and even fewer women among the scientific elite. Currently in the United States, only about 12 percent of science Ph.D.'s are women. Women comprise only 3 percent of the 1970 Ph.D.'s in physics and astronomy; 8 percent in chemistry; about 15 percent in the biological sciences; 18 percent in sociology; and 24 percent in psychology.[4]

This pattern has been remarkably consistent over the past fifty years. A recent National Academy of Sciences report notes:

> Women receive 40 percent of the baccalaureate degrees granted in the United States, and the percentage is increasing. They receive 32 percent of the master's degrees, but this proportion has remained constant for many years. The percentage of women among United States doctorate recipients dropped from 15 to 9 percent between 1920 and 1950, but a gradual increase restored the value to 11 percent by 1960. Since 1960, the proportion of women among all doctorate recipients has remained constant, not only for the total, but for each summary field.[5]

Looking at the rolls of the National Academy of Sciences and the list of Nobel laureates, we can see that there are very few eminent women scientists. Only 8 of the 866 members of the National Academy are women, and only 5 of the 281 scientists who have received the Nobel Prize have been women; of these five, Marie Curie, Irene Joliot Curie, and Gerty Cori shared their prize

4. *Summary Report, 1967 Doctorate Recipients from United States Universities,* National Research Council, National Academy of Sciences (Prepared in the Research Division of the Office of Scientific Personnel, OSP-RD-1, May 1968); *Summary Report, 1968 Doctorate Recipients from United States Universities,* National Research Council (Prepared in the Education and Employment Section, Manpower Studies Branch, Office of Scientific Personnel, OSP-MS-2, April, 1969).

5. Doctorate Recipients from United States Universities 1958–1966, Publication 1489, National Academy of Sciences (Washington, D.C. 1967), p. 107.

with their husbands.[6] How can we account for these facts? Has science, which otherwise has a strikingly good record in approximating the ideal of universalism, systematically excluded women from its ranks?

There is widespread belief among scientists and nonscientists that women are discriminated against in appointment to high-prestige science departments; that they are less likely than men to receive promotions to tenured positions; must wait longer for their ascendency; are less often the recipients of honorific forms of recognition; and that their salaries are lower at every academic level. Only recently, however, have groups of women become vocal in their protest against the apparent injustices against women scientists. We need look no further than recent issues of *Science* for stormy letters of complaint from women scientists who blast the "community" for out-and-out discrimination.[7] One irate female biologist put it this way:

> The physical scientists keep their female graduate students down to a tiny minority, and thus avoid the paradox of training those women they would not employ. Biologists, however, have not been so fastidious. Beguiled by lavish federal funds for graduate student training the men who run our graduate biology departments have played a shameless numbers game, eagerly enticing women students to swell department rolls and bring in money. The inconvenient fact that winning a Ph.D. would be unlikely to entitle a girl to more than second-class citizenship in the scientific world has, of course, gone unnoticed. The acceptance of this situation is to many men merely a matter of being realistic, though some are honest enough to admit that "cynical" would be a better word. The best word for it would be stronger.[8]

Is this perception of the operation of the reward system of science accurate? Here we shall examine data to determine whether apparent discrimination in science is actual discrimination. It is not without import for our abiding concern with universalism in science to assess the extent to which women are treated fairly once they become scientists. If all or some of the commonly held beliefs about

6. Harriet A. Zuckerman, "Women and Blacks in American Science: The Principle of the Double Penalty" (Paper presented at the Symposium on Women and Minority groups in American Science and Engineering, California Institute of Technology, 8 December, 1971), pp. 34–35.

7. Among the recent issues of *Science* which have had letters and papers on the place of women in science, see 7 May, 1971; 16 April, 1971; 16 July, 1971.

8. *Science,* 7 May 1971, p. 514.

women in science turn out to be empirically correct, we can conclude that, at least in terms of sex status, the ideals of rationality and universalism in science are being compromised. We consider here only "survivors" among women scientists—those who have overcome the cultural forces impeding their choice of science as a career and have actually received their doctorates. In order to assess the degree of discrimination against women, it is necessary to compare the careers of women scientists with those of their male colleagues. We want to estimate the influence of sex status on recognition, while controlling for the type of training the scientists have received and the quality and quantity of their scientific research.

To this end, we collected data from several sources on a sample of 499 men and women scientists in the fields of chemistry, biology, and psychology. This sample was generated by selecting men matched to a subsample of 1,547 women doctorates studied by Helen Astin.[9] Astin studied 1,979 women who received their doctorates from American universities in 1957 and 1958. Data on this population were collected through a mail survey in 1965. Eighty-three percent of all 1957 and 1958 women doctorates, or 1,547, responded to the survey. We have selected male matches for only the 749 women who received doctorates in the physical, biological, and social sciences. These male matches were drawn from the doctorate record file of the Office of Scientific Personnel (OSP), which includes data on almost all doctorate recipients in the United States.[10] Four matching criteria were used: year of doctorate, university where the Ph.D. was earned, field, and specialty.[11] Here we

9. We thank Dr. Helen Astin for the help that she has given us in this study. She has provided us with the basic data set and codebook for the women scientists. Without her assistance this chapter would not have been attempted. Her own results for women scientists are reported fully in *The Woman Doctorate in America* (New York: Russell Sage, 1969).

10. Astin's study also made extensive use of the doctoral record file of the Office of Scientific Personnel. We thank Dr. William Kelley and Ms. Clarebeth Maguire of the OSP for their help in obtaining the basic matches used in this study.

11. Drawing the sample involved a number of contingency operations. First, priority was given to matching on university department, then on field, on specialty, and on year. If, for example, we were able to generate a match for a woman in biochemistry at the University of Missouri, but there was no Ph.D. in 1957, we searched for a match in 1958. If a match was found in 1958, we took this "adjusted" match. If we had to, we selected a man who received his degree from the same university, in the same field, in 1957. Finally, we generated two male matches for every woman in the sample. This was done to ensure that we got a high proportion of the men with complete data. Since our primary source for background information on the male

report data only on the 499 men and women scientists who were academically employed in chemistry, biology, or psychology departments. Career data for the women were obtained from Astin, and data on selected social characteristics and on social mobility were collected for the entire sample from *American Men of Science*.[12] Publication counts for the years 1957 through 1969 were gathered from the appropriate abstracting journal.[13] Citation data

---

scientists was *American Men of Science,* if the first match was not found in the *AMS,* we selected the second match, if he was found in the *AMS.* Where both men were not to be found in the compilation, we selected the first match. Both male matches were, in every case, randomly selected from the available matches.

It turned out that, in about 19 percent of the cases, we used the second match. There are at least two reasons why this procedure is unlikely to distort the results presented here. First, most of the male physical and biological scientists were found in the *AMS;* a much smaller proportion of the social scientists could be found in the edition of the *Behavioral and Social Science* volumes of *AMS.* Certainly the results for chemistry and biology are unlikely to be distorted at all; psychology could have some biased results, if one assumes that the men in *AMS* are more eminent than the women to whom they are matched. Second, in many cases, the second male match was linked to a woman who was in the *AMS.* Consequently, by selecting the second match in this limited number of cases, we are actually controlling somewhat for "minimal" eminence. We are not considering data for physics and mathematics because there are so few women doctorates in these fields that any separate analysis by field would involve a very small number of women scientists. Other social sciences and humanities were not included because we could not obtain reliable data on the publications and citations to academics in these fields. Further, we have excluded from analysis here nonacademic scientists, because the reward system in industry and government science are far more varied than in academic science. Since we wanted to consider a more or less uniform reward structure, we limited analysis to academics.

For this sample of men and women scientists within the three fields under consideration, women are more likely than men to hold academic positions. In chemistry, 55 percent of the women held academic positions compared with 31 percent of the men. Women in chemistry were also slightly more likely than men to hold positions in universities. The same pattern was observed for the biological sciences, where 74 percent of the women and 68 percent of the men were located in academic positions. Only in psychology was the proportion of men in academics greater than the proportion of women: 52 percent to 38 percent. While women are slightly more likely than men to hold positions at colleges, they are not significantly underrepresented compared to men at university settings.

It should be emphasized that after the initial matching procedure, the original pairings were not maintained during analysis. The entire sample was treated in aggregate form during statistical analysis. Little distortion appears to have crept into the sample, however. Among academic men and women scientists, the correlation between sex status and prestige rank of doctoral department was still virtually zero.

12. We use the eleventh edition of the *American Men of Science.*

13. We used the following sets of abstracts for the publication counts: *Science Abstracts, Chemical Abstracts, Biological Abstracts,* and *Psychology Abstracts.* Both single-author and multi-author papers were counted.

were collected from the *SCI*.[14] Finally, the measured intelligence and high school academic records were obtained from the Office of Scientific Personnel.

We want to know the extent to which men and women scientists receive recognition. We shall consider once more two types of recognition: positional success and reputational success. As an indicator of positional success, we will use the prestige rank of the scientist's academic department as measured by Cartter's American Council of Education study. Universities not included in Cartter's study were rated lower in prestige than those that were ranked, and all colleges were scored lower than universities. While this may involve occasional inaccuracies, universities are, in general, the locus of research activities, and a substantial proportion of the eminent scientists in any era can be found at the better large universities. An additional indicator of positional success that we consider is academic rank, although rank must obviously be coupled with prestige of the academic department before it is a meaningful indicator of success. Unfortunately, for this sample of scientists we have only limited data on reputational success; we do not have data on the visibility or perceived quality of work of the men and women scientists. We do have data on the number of honorific awards that these scientists have received. We use this as an estimate of reputation.

Sex status and prestige of academic affiliation

Until recently, there have been few inquiries into the comparative positions held by men and women scientists. Those which have been done suggest that women tend to occupy lower-prestige positions than male peers. Budner and Meyer, looking at a sample of social scientists, found that among those at universities, women were more likely than men to be at lower-prestige institutions.[15] They found 22 percent of the women at "high" quality schools compared with 38 percent of the men. Correlatively, 55 percent of the women and

14. Citation data were collected for six years: 1961, 1964, 1965, 1967, 1969, 1970. If we had only wanted a single measure of the "quality" of scientific work produced by the scientists, we undoubtedly could have taken counts for only one year. Extensive citation data were collected in order to examine the differences between the impact of early and late work for men and women scientists. In all of these counts, we have excluded all self-citations.

15. Stanley Budner and John Meyer, "Women Professors," as reported in Jesse Bernard, *Academic Women* (University Park, Pennsylvania: The Pennsylvania State University Press, 1964). Budner and Meyer present data drawn from Paul F. Lazarsfeld and Wagner Thielens, Jr., *The Academic Mind* (Glencoe, Illinois: The Free Press, 1958). They used the Berelson ratings of the quality of a university.

30 percent of the men were located at either "medium-low" or "low" quality institutions.[16] Rossi, studying the distribution of men and women sociologists, also found a relationship between sex status and the prestige of the department affiliation of scientists.[17] Both of these studies did not control for age or the type of training received by the scientists.

Our data, which do match men and women scientists for professional age and place of training, allow us to determine more adequately the relationship between positional success and sex status. We begin with an apparently striking datum: the zero-order correlation between sex status and prestige of academic department is quite small ($r = -.07$). This suggests that men and women scientists with similar training tend to wind up in academic departments of equal prestige.[18] This zero-order correlation coefficient, however, conceals some variation between the three academic fields making up the sample. In chemistry and psychology, women are somewhat less likely than men to be found in top departments. The zero-order correlation for the two fields is $r = -.27$ and $r = -.23$, respectively. In biology, where there is a greater number of women in academic positions than in the other two fields, women do slightly better than men ($r = .06$).

Academic rank

Promotion to the senior ranks of associate and full professor is a common aspiration of all academicians. Intense competition exists for promotion to the higher ranks, especially at prestigious departments. Scientists are not only concerned with achieving high rank, but in doing so as quickly as possible. Some measure of recognition is garnered by the distinction of being a "young" associate or a "young" full professor.

The literature dealing with the academic rank of men and women scientists shows inconsistent findings. Perrucci, for example, found that women tend to hold lower ranks than their male peers.[19] She

16. Budner and Meyer, "Women Professors," p. 93.

17. Alice A. Rossi, "Status of Women in Graduate Departments of Sociology," *The American Sociologist* 5 (February, 1970): 1–12; see also idem, "Equality Between the Sexes: An Immodest Proposal," *Daedalus* (1964): 98–143; idem, "Women in Science: Why So Few?" *Science*, (28 May 1965); 1196–1202.

18. For this analysis, sex was coded: 1 = men, 2 = women. Thus a negative sign before a correlation indicates that women are less likely than men to be in high ranked departments, or less likely to hold high academic rank.

19. Carolyn Cummings Perrucci, "Minority Status and the Pursuit of Professional Careers: Women in Science and Engineering," *Social Forces* 49 (December 1970): 245–59.

found that men were more likely to have tenure than women. Simon et al. found that women in all fields are less likely than men to hold the ranks of associate or full professor.[20] They found that marital status for women could specify the relationship between sex and rank: unmarried women were about as likely to hold high-ranked positions as men. Other studies, particularly that of Bayer and Astin, show no significant differences in the ranks of men and women academics in either college or university settings.[21] Each of these studies presents problems in interpretation since age, educational background, and types of institutional affiliation are not always adequately controlled, and the number of women sampled is often extraordinarily small.

Since all the scientists in our sample received their Ph.D.'s at the same time (1957–58) the correlation between sex status and academic rank may be used as an indicator of the extent of recognition of the men and women scientists. The zero-order correlation between sex status and academic rank for the entire sample is −.29. There are field differences, ranging from an insignificant difference in the ranks of men and women chemists ($r = -.05$) to substantial differences among biologists ($r = -.34$) and psychologists ($r = -.34$). As noted above, academic rank is a truly meaningful indicator of recognition only when it is coupled with the prestige of the department at which high rank is achieved. The facts of academic life are clear: if one stays in the business long enough, one is bound to be promoted. The critical part of recognition associated with rank is the achievement of high rank in a high-prestige department at a relatively early age. In fact, Hargens has suggested that one mechanism by which the social system of science handles failure is by promoting scientists in rank while demoting them in terms of the prestige of the department with which they are affiliated.[22] Evidence that longevity alone will produce changes in academic rank can be seen in the almost total absence of a correlation between productivity and academic rank within less distinguished academic departments ($r = .04$). At better departments there is a significant correlation, as we would

20. Rita James Simon, Shirley Merritt Clark, and Kathleen Galway, "The Woman Ph.D.: A Recent Profile," *Social Problems* 15 (Fall 1967):221–36, especially pp. 228–29.

21. Alan E. Bayer and Helen A. Astin, "Sex differences in Academic Rank and Salary among Science Doctorates in Teaching," *Journal of Human Resources* 3 (1968): 191–200.

22. Lowell Hargens, "The Social Contexts of Scientific Research" (Ph.D. diss. University of Wisconsin, 1971).

expect, between academic rank and the quantity of scientific output (r = .24). When we divide our sample into two groups, departments rated in the Cartter study as distinguished, strong, and good, and all other university and college departments, we still find significant zero-order correlations between sex status and academic rank. In the better departments, the correlation is −.35; in the less distinguished departments r = −.21.

Sex status and salaries

A widely held belief among women scientists is that their salaries are significantly lower than men's, even after controlling for type of affiliation, academic rank, and length of tenure. Data from a variety of sources tend to confirm this belief.[23] These studies, using different sampling techniques and focusing on different scientific disciplines, arrive at slightly different estimates of the differentials between the salaries of men and women scientists. However, the pattern of the findings is consistent. Bayer and Astin, for instance, found that after five or six years of employment, women in colleges and universities earn about 92 percent of the salaries of men in both the natural and social sciences.[24] As expected, salary differentials are greater among senior- as compared to junior-rank scientists. Our data did not include the salaries of the male scientists. Thus we cannot estimate the differentials in salaries for the matched sample.

Honorific awards

There are only limited data available on the honorific recognition granted to women scientists. We have already noted their virtual absence from the elite academies of science and their low representation among Nobelists. Simon's data, however, suggest that women are more likely than men to receive postdoctoral fellowships and memberships in honorary societies.[25] For our sample of scientists, we counted the total number of honorific awards and postdoctoral fellowships listed after their names in *American Men of Science*. For the entire sample, the correlation between sex status and number of honorific awards was −.07, men being slightly

23. See, among others, Sylvia Fleis Fava, "The Status of Women in Professional Sociology," *American Sociological Review* 25 (April 1960):271–76; Bayer and Astin, "Science Doctorates"; Michael A. LaSorte, "Sex Differences in Salary Among Academic Sociology Teachers," *American Sociologist* 6 (November 1971): 304–7; Gloria Lubkin, "Women in Physics," *Physics Today* 24 (April 1971); David A. H. Roethal, "Starting Salaries—1970," *Clinical and Engineering News* (November 23, 1970).

24. Bayer and Astin, "Science Doctorates," p. 196.

25. Simon et al., "The Woman Ph.D.," pp. 232–33.

more likely to have received awards. Once again, we observed field differences: the zero-order correlation for chemistry was −.18; for biology r = −.05; for psychology r = −.14. In summary, there are only small differences in the level of reputational success of men and women scientists in our matched sample.

We have presented a set of zero-order correlations and evidence from the literature which focus on the relationship between sex status and various forms of recognition (for a summary, see table 1). What can we conclude from these data? The data indicate that in the fields we have examined, men generally do slightly better than women in the prestige rank of their academic department, their academic rank, the receipt of honorific awards, and their average salaries. On the basis of observed zero-order correlations such as these, some women scientists believe that discrimination is operating in the scientific community. Among Astin's sample of women scientists, 25 percent cited experiences of discrimination in hiring practices; 40 percent had experienced differentials in salaries; and 33 percent cited differentials in policies on tenure and promotion.[26] But the fact is that we cannot conclude on the basis of the data presented thus far that women are being discriminated against. If the differences in recognition received by men and women scientists were a result of differential role-performance, then there would be no discrimination and the social system would be operating in accordance with its universalistic and rational ideals.

Table 1. Zero-order correlations between sex status and three forms of scientific recognition

	All Fields	N	Chemistry	N	Biology	N	Psychology	N
Rank of department	−.07	320	−.27	38	.06	184	−.22	98
Academic rank	−.29	413	−.05	53	−.34	248	−.34	112
Number of honorific awards	−.07	379	−.18	40	−.05	236	−.14	103

NOTE: There were missing data for this sample. We have used pair-wise correlations. The "N" columns give the number of cases on which the Pearson correlation is based. Sex status was coded: 1 = male; 2 = female.

But before we consider the influence of differential role perform-

26. Astin, *The Woman Doctorate*, p. 106.

ance on the distribution of rewards, we must first entertain the possibility that sex status is, in fact, not a functionally irrelevant status for the performance of scientific roles. Perhaps male scientists have more native ability than female scientists. Although we have previously pointed out the inadequacy of I.Q. tests as a measure of scientific ability, it is the only measure that is available, and provides us with crude measures of natural aptitude.

The data do not support the hypothesis that differences in the level of recognition result from differences in the innate ability of the men and women Ph.D.'s. Data presented by Harmon have suggested that the measured intelligence of women doctorates is on the average slightly, if not significantly, higher than that of male doctorates.[27] Our analysis of a subset of Harmon's data suggests that within the three fields we studied, and at every level of doctoral department prestige, women doctorates have on the average slightly higher I.Q.'s than their male colleagues. While these differences are not always statistically significant, they are striking in their uniformity.[28] In short, if we are willing to accept I.Q. as a rough indicator of general innate ability, women and men start off on at least equal footing. On this basis, we would expect women scientists to fare just as well as men.

Since we have demonstrated that rank in the scientific stratification system depends heavily on the quality of published research—and indeed on its production at a reasonable rate—we must examine the productivity patterns of men and women scientists. How, if at all, do the publication patterns of men and women scientists differ?

PRODUCTIVITY OF MEN AND WOMEN SCIENTISTS

On the whole, the sample of men and women scientists was not extraordinarily prolific, but by no means below normal, in research output. The mean "lifetime" productivity for the 499 academic scientists was nine papers. However, few scientists in the sample published much in any given year. For example, in 1959, one or two years after receipt of the doctorate—usually prolific years for young scientists—53 percent of the sample failed to publish a single paper,

27. Lindsey R. Harmon, "High School Ability Patterns: A Backward Look from the Doctorate," Scientific Manpower Report No. 6 (Washington: National Research Council, 1965).

28. For a detailed discussion of the relationship between I.Q. and sex status, see Jonathan R. Cole, "American Men and Women of Science" (Paper presented at the annual meeting of the American Sociological Association, Denver, Colorado, August, 1971).

and 34 percent published just one. In most years, between 70 and 80 percent of the scientists published nothing.

But what is the correlation between sex status and productivity?[29] It turns out that male scientists in the matched sample, on the average, are more productive than women. This is clearly reflected in the zero-order correlation for the entire sample between sex status and "lifetime" productivity ($r = -.36$)[30] (see table 2). The differential in publication rates remains fairly constant over the entire career of men and women scientists, although the initial differences in productivity two years after receiving the doctorate are insignificant ($r = -.06$). Publication activity in the years immediately following the doctorate is a better predictor of later productivity rates for men than it is for women.[31]

There are numerous alternative explanations for the lower productivity of women scientists. Consider only two which are frequently offered. First, it is claimed that the lower publication rates of women result from family obligations which prevent them from spending as much time working as men. Since we had data on the marital status and number of children for both the men and women scientists, we could test this hypothesis. Productivity was regressed on three independent variables—sex status, marital status, and number of children. Sex differences in productivity persisted. The net effect of sex status on productivity after controlling for these family statuses is essentially equivalent to the zero-order correlation ($b^*_{PS \cdot MF} = -.35$).[32]

But surely for women, marital status and family size has a qualitatively different influence on professional life than for male scientists.

29. The total number of published papers, both single-author and collaborative, was used as the measure of productivity. "Lifetime" productivity here refers to the total number of published papers between 1958 and 1969; "early" productivity includes just the years 1959–65; "later" productivity refers to the years 1966–69.

30. The zero-order correlation between "lifetime" published productivity and total productivity prior to 1965 is extremely high ($r = .95$).

31. In all fields the correlation between early and later productivity is higher for men than for women. For a detailed discussion of this point, see J. R. Cole, "American Men and Women of Science."

32. In presenting the partial regression coefficients, we use the following subscripts: S = sex status; P = productivity; Q = quality of research; M = marital status; F = family size; R = rank of current department (1965); A = number of honorific awards and postdoctoral fellowships; B = academic rank; C = college affiliation; U = university affiliation. The failure of family status to reduce the correlation between sex status and productivity results from the small zero-order correlation between productivity and marital status ($r = .11$), that is, married scientists publishing slightly more than unmarried ones) and nonexistent correlation between productivity and family status ($r = .01$).

Table 2. Zero-order correlations between sex status and various indicators of scientific output

	Entire sample of scientists	N	Chemistry	N	Biology	N	Psychology	N
Productivity (Total number of papers)								
First eight years (1958–65)	−.37	495	−.38	59	−.37	301	−.34	135
Total career (1958–69)	−.36	495	−.38	59	−.37	301	−.35	135
Quality of work								
First eight years (Total citations: 1961, 1964, 1965)	−.19	362	−.27	59	−.18	303	*	
Total career (Total citations: 1961, 1964, 1965, 1967, 1969, 1970)	−.24	497	−.37	59	−.24	303	−.28	135

* Citation data for most psychology journals were not abstracted by SCI during the early 1960s. Thus we collected citation data for only 1969 and 1970 for the psychologists. However, since the correlation between the total number of citations scientists' receive in the 1961, 1964, 1965 index is highly correlated ($r = .93$) with the total number of citations, the use of later citations as an indicator of earlier use of work will not involve gross error.

In short, regression results may conceal interaction between sex status and family statuses. To test for interaction, the sample was first divided into men and women and then further divided into those who were unmarried, married without children, those with one or two children, and those with three or more children. The mean number of "lifetime" publications was computed for each of the eight subgroups. The data in table 3 testify that sex differences have a much greater influence on publication patterns than do family statuses. Consider the following striking fact: unmarried women scientists publish far less than men scientists in all family categories. We may conclude that family status cannot account for the differential rates of productivity of men and women scientists. Family obligations only influence the productivity rates of women having three or more children. Women with smaller families are no less likely to publish than unmarried women.

Table 3. Mean number of "lifetime" publications for men and women scientists, controlling marital status and family size

	Mean Number of Publications (1958–69)				
		Marital and family status			
Sex	Total	Unmarried	Married with no children	One or two children	Three or more children
Men	13 (254)	10 (30)	13 (30)	14 (109)	12 (85)
Women	5 (241)	6 (62)	8 (33)	6 (83)	3 (63)

Another explanation offered for the lower productivity of women is their occupational location. Women are more frequently found in college settings, where research is not the norm or a prerequisite for promotion. Astin summarizes the hypothesis:

> Studies indicate that institutional affiliation (college or university) accounts for most of the differences observed [in productivity]. Essentially, academic women have a greater tendency to work in colleges, while academic men are employed by universities. Since university-employed persons publish more than college-employed persons, regardless of sex, the greater overall productivity of academic men is self-explanatory.[33]

33. Astin, *The Woman Doctorate*, p. 85. Although Astin asserts that academic affiliation can account for differentials in productivity, she does not present any data to support this hypothesis.

Our data enabled us to test this hypothesis. We regressed productivity on institutional affiliation and sex status. Although scientists at universities are in general more prolific and produce work of greater consequence than those located at colleges, the data suggest that institutional affiliation fails to even slightly modify the relationship between sex status and scientific output. The partial regression coefficient between sex status and productivity is virtually equal to the zero-order correlation between the two variables ($b^{*}_{PS \cdot CU} = -.28$). As a further test of this hypothesis, we computed the correlation between sex status and productivity only, for scientists working at universities. The correlation was −.35, the same as we found for the entire sample. In short, men at universities are more productive than their female colleagues at the same location. The same pattern is observed at college settings.

We might expect that in better science departments, the correlation between sex status and published productivity would be attenuated. This turns out not to be the case. When we divided the scientists into those located in 1965 at better academic departments (distinguished, strong, and good by the Cartter ratings) and those at less prestigious departments, we found a zero-order correlation of −.31 between sex and productivity at the better departments and a correlation of −.25 at comparatively poorer departments. No matter how the sample of academic scientists was divided, productivity differentials between men and women scientists persisted.

While there surely are other variables, such as differentials in teaching responsibilities, access to research funds, and access to collaboration with other outstanding scientists, which might account for the differences in published productivity of men and women and which remain untested here for lack of data, our efforts to explain different rates of productivity with variables at hand proved unsuccessful. In order to fully account for the correlation between sex status and the rate of scientific output, it may be necessary to look outside the institutional structure of science; to examine carefully the prior experience and socialization of women in the larger society which influences their performance after they enter science.

Although women publish fewer scientific papers than men, it is possible that the papers that they do publish may be of higher quality. Perhaps because women, in general, are under less pressure to "succeed," they may be more likely to be "perfectionists." The data suggest that women are not, in fact, as highly cited as men scientists ($r = -.24$). In the six years for which we have citation data, papers published by men received a mean of 11 citations per year and those by women a mean of 4 per year. In all three scientific fields women receive fewer citations per published paper than men.[34]

Let us pause and recapitulate. We have shown that women receive slightly less recognition than men, that they have the same or higher I.Q.'s than men, and that they produce fewer and less frequently cited papers. To infer from simple zero-order correlations between sex status and recognition that women are being discriminated against, would be to oversimplify reality. Such inferences are often misleading, for they see social behavior as two-dimensional. The task at hand is to bring into clear perspective the multivariate aspects of social reality affecting men and women in the social system of science. Therefore, we must now look at the relationships between sex status and forms of recognition, controlling for differential role-performance.

Although we cannot explain differences in productivity of men and women scientists, substantial variability exists in the productivity of women scientists. The real issue is whether we can explain differential rewards by differential role-performance. In short, when we control for quality and quantity of research output, do men still receive greater recognition than women?

SEX STATUS AND RECOGNITION, CONTROLLING FOR SCIENTIFIC ROLE-PERFORMANCE

Recall that for the entire sample of academic men and women in three fields, sex status was only slightly correlated with rank of department ($r = -.07$). When we now control for the level of productivity of the scientists, the zero-order correlation between sex status and rank of current department disappears ($b^*_{DS.PG} = .02$). The zero-order correlations between sex and department prestige within the three specialties varied ($r = -.27$ in chemistry, $r = -.23$ in psychology, $r = .06$ in biology). The fact that there are different zero-order correlations in the three fields does not imply that there is less "discrimination" in one field or another. Take the case of biology, where women in the matched sample are more likely to be found at high-prestige departments than their male peers. Women biologists could still be the objects of discrimination. For if women

34. The correlation between early and later citations are higher for men than for women in both chemistry and biology, the two fields for which we have citation data over time.

biologists are on average more productive and produce higher quality work than their male peers, then it is possible that they are still not being rewarded in direct relation to their performance.

When we regress department prestige on sex status, productivity, and quality of research, the association between sex status and rank of department is reduced significantly in both chemistry and psychology. In chemistry, the partial regression coefficient for sex status, after controlling for quantity and quality of research, is $b^*_{RS.QP} = -.13$. As expected, scientific output is a more influential factor in determining a scientist's rank of department than is sex status. The net effect of research quality is $b^*_{RQ.SP} = .23$; of research quantity $b^*_{RP.QS} = .19$. For psychology, the zero-order correlation is reduced even more substantially after controlling for scientific output ($b^*_{RS.QP} = -.07$). Here too, quality of research is a more significant influence on rank of department ($b^*_{RQ.SP} = .36$), and quantity has only slight independent effect ($b^*_{RP.SQ} = .11$). In the biological sciences, the pattern is different. When we control for the bulk and quality of work, women are still slightly more likely to be found at higher-ranked departments. Indeed, the partial regression coefficient increased slightly, $b^*_{RS.QP} = .13$. The results are summarized in table 4A.

While the regression coefficients suggest that women psychologists and chemists are slightly less likely than their male peers to be located at the prestigious departments, independent of the quality and quantity of their scientific output, three qualifications must be considered. First, the standard errors of the regression estimates for both chemistry and psychology are greater than half the beta coefficients, suggesting the possibility that this finding may, in fact, be a statistical artifact of random errors in the data. Second, in all fields "sex" independently explains an extremely small portion of the total amount of variance on rank of department. In fact, the coefficient of determination between rank of current department and the three independent variables for chemistry is only $R/r = .44$, suggesting that other relevant factors such as where the doctorate was earned, or perhaps functionally irrelevant criteria other than sex status, influence the rank of the scientist's department. Finally, since holding a position at any department involves a strong component of choice, women may for a variety of reasons be self-selecting themselves out of appointments at better departments.[35] This is

35. Differentials in rewards to women may well be a consequence of conscious choice. For example, many women choose to drop out of science for a limited amount

generally consistent with the finding in chapter four, where we noted that rank of department had a lower correlation with both quality and quantity of output than other less "voluntaristic" forms of recognition.

In summary, while indicating very limited independent effects of sex status on the prestige of department in two of the three disciplines, the data suggest that the influence is not substantial. When

Table 4. Zero-order correlations and regression coefficients in standard form for relationship between sex status and three forms of scientific recognition for three scientific fields

Form of recognition	Correlation between sex status and form of recognition	Betas between sex status and recognition, controlling for quality and quantity of research	
A. *Rank of current department*	*r*	*b**	*N*
Total sample	−.07	.02	320
Chemistry	−.27	−.13	38
Biology	.06	.13	184
Psychology	−.22	−.07	98
B. *Academic rank*			
Total sample	−.29	−.29	413
Chemistry	−.05	−.06	53
Biology	−.34	−.36	248
Psychology	−.34	−.31	112
C. *Number of honorific awards*			
Total sample	−.07	.00	379
Chemistry	−.18	.00	40
Biology	−.05	.00	236
Psychology	−.14	−.07	103

we turn to the issue of advancement to higher academic ranks, the story becomes more complex. The zero-order correlations and standardized partial regression coefficients for the entire sample and each of the three fields are presented in table 4B. Controlling for

of time to begin families. Moreover, once an individual leaves science for even a limited amount of time, or works with only half-energy, it becomes increasingly difficult to keep pace with advances in scientific knowledge.

quantity and quality of scientific output does not reduce the relationship between sex status and academic rank. At each level of productivity, women are less likely to receive promotions than male scientists. This result correctly suggests, of course, that productivity ($r = .07$) and quality of output ($r = .03$) are virtually uncorrelated with academic rank.

Thus far, these findings skirt the important issue of the relationship between sex status and promotion at specific types of departments. Are women promoted at the same rate as men at the better and lesser academic departments? To answer this question, we once again divided the sample into those men and women at more and less prestigious departments. One group consisted of scientists at "distinguished," "strong," and "good" departments; the second group included those at all other university departments and colleges. For each group separately, we regressed academic rank on sex status, productivity, and quality of scientific output. At both types of department, sex status is still independently related to academic rank.[36] This is to be expected since quality and quantity of research is not correlated at all with rank at the lesser departments and is only moderately correlated with rank at the better departments. At this juncture, we must conclude that there is some influence of sex status on the determination of academic rank within the scientific community.

The association between sex status and academic rank is not a statistical artifact. At least for academic rank, some measure of sex-related particularism apparently operates within the reward system. But even here we must proceed with caution. We know that the key determinant of academic rank is length of professional experience. If some of the women in our sample had at some point taken leaves of absence, this might explain the findings of table 4B. Although these results are based only upon those women who were working full-time in 1965, we do not have data on whether the women were continuously in the labor force for the entire eight-year period following their degrees. We cannot, for example, determine which of these women took leaves of absence at some point during the first eight years of their careers. Astin reports: "Of the 1,214 women

36. Within distinguished, strong, and good departments, the zero-order correlation between sex status and academic rank is $r = -.35$; the partial regression coefficient between sex status and academic rank, controlling for quantity and quality of output, is $b^*_{BS.QP} = -.31$. For all other departments the correlation between sex status and academic rank is $r = -.21$; the partial regression coefficient is identical to the zero-order correlation.

doctorates who were fully employed at the time of the survey, 957 (79 percent) had never interrupted their careers; 18 percent reported career interruptions lasting from 11 to 15 months, with a median period of 14 months."[37] Since our sample of women is a subsample of Astin's and since we are considering promotions seven years after the doctorate (a particularly important period for mobility of academics), these career interruptions of women could influence their rate of promotion. In fact, when Bayer and Astin took this employment pattern of women into account, they found only minor differences in the rate of promotion of men and women in the sciences.[38] In order to correct for different lengths of time in the labor force, they compared women who earned their Ph.D.'s in 1957–58 with men who earned their doctorates in 1958–59. They found that, in this group, sex had no influence on academic rank.

One group of women with a high probability of taking some time off from their careers are those who have young children. We would accordingly expect that women with preschool-age children and preteen-age children would be less likely to receive promotions, since they either have had less time to spend on their careers or have actually taken leaves of absence. The data suggest that this is the case. We divided the sample into women who had appointments at higher-prestige universities and those in less prestigious universities and colleges. At the high-prestige institutions, the correlation for women between number of preschool children and academic rank was $r = -.36$; the correlation was even stronger between number of preteen-age children and academic rank ($r = -.49$). These data point to the difficulty in locating the reasons for the correlation between sex and academic rank. Is it primarily due to particularism operating within science, or to factors external to the scientific community? At lower-ranked departments, where promotions do not depend as much on publication records, the zero-order correlation between number of preschool children and academic rank is somewhat lower ($r = -.18$). There is no correlation between number of preteen-age children and academic rank at the less distinguished settings ($r = -.01$).

Let us turn now to the one indicator of honorific recognition for which we have some empirical data: the receipt of honorific awards and postdoctoral fellowships. For the sample of some 500 men and women scientists in three fields, we found a small correlation be-

37. Astin, *The Woman Doctorate,* p. 58.
38. Bayer and Astin, "Science Doctorates."

tween sex status and number of honorific awards ($r = -.07$). When we control for the quality and quantity of published research, the influence of sex status on this form of recognition is completely eliminated. The partial regression coefficients are presented in table 4C. Again, high productivity and quality of research are the primary determinants of honorific recognition. We may tentatively conclude that there is virtually no difference in the recognition of men and women scientists in terms of the awards that they are likely to receive.

Unfortunately, we do not have salary data for the men in our sample. It is therefore impossible for us to explain differences in the salaries of men and women scientists. However, we must point out that any adequate test for discrimination must take into account differences in the role-performance of men and women. Without adequate controls for the actual performance of scientific roles, differences in salaries are not subject to easy interpretation.

In brief summary, the data that we have presented suggest that men and women in American science with similar backgrounds are treated similarly by the reward system of science. There does appear to be some differential treatment of women in academic promotion—at least differences which could not be explained by the measures of role-performance that we have at hand. But women are not significantly underrewarded in honorific recognition or in their academic affiliation, once we take into account their achievements as scientific researchers.

We have been moving toward a specification of the conditions under which sex status influences recognition granted to scientists. We have focused primarily on whether differences in productivity and quality of work explain the slightly unequal treatment women receive in science. It is possible that these regression procedures are masking a specification of the findings. We thought it possible that totally unproductive women might be treated differently than totally unproductive men. Functionally irrelevant characteristics are likely to be activated when there are no or few functionally relevant criteria on which to judge individual performance. They will also be imported in situations where there is only limited agreement as to what are the relevant criteria for judgment. In science, one basic criterion for judging role-performance is the quality and, to a far lesser extent, the quantity of research produced by a scientist. Of course, if two scientists being considered for a job have both failed to publish anything, then quality of work becomes a meaningless criterion. At the best science departments virtually every member publishes scientific papers. But at lower-prestige colleges, far removed from the hub of scientific activity, there often is a majority of a faculty that publishes no research papers. We hypothesize that sex status would be more likely to be imported into a social situation in which scientists published little or nothing than in one containing very productive scientists. In fact, in the absence of published papers to testify to the seriousness of the scientist, some "status judges" may well contend that sex is a relevant basis for awarding positions to scientists. These arbiters argue that women are much greater risks than men, that they are more likely to withdraw from science for a variety of reasons, and therefore should not be hired in the first place.

To test these hypotheses, we would ideally search for science departments in which no one published research. The data in its present form could not be arranged for such a test. Therefore, we decided to examine briefly the career patterns of one extreme group: those truly "silent" scientists who had not produced a single scientific paper during the first four years in their careers yet had not dropped out of science.

Among silent scientists, does "Susan the silent" suffer in comparison with "Stanley the silent"? Evidence substantiating such a pattern would suggest, although hardly prove, that functionally irrelevant statuses have been activated; that sex differentials obtain. Our data allow us to make a first test of this idea. Consider appointment to departments of varying prestige. Are silent women less likely than their male peers to have jobs at distinguished, strong, good, or even adequate departments? The data imply that this is not the pattern, and suggest that we cannot accept the hypothesis. The zero-order correlation between sex status and prestige rank of a scientist's 1965 academic affiliation for the forty-eight silent scientists is $r = .15$. This correlation is not statistically significant. However, it suggests that women who do not publish any papers are slightly more likely to have jobs at more prestigious departments than silent male scientists. Clearly, these empirical findings just scratch the surface of an adequate test of these theoretical ideas. However, they suggest that science treats its totally unproductive men and women equally well in this one form of recognition.

An emerging motif in this book is that of "accumulative advantage." We have suggested that scientists who are placed in structurally advantageous positions as a result of outstanding role-performance are given certain advantages by virtue of attaining these positions. There are accumulating effects of being at an outstanding science department, for example, in terms of future positional and reputational success. There is, however, a second analytic and darker side of the process of accumulation of rewards in science. It is the accumulation of failures—the process of "accumulative disadvantage."[39]

We have looked at the recognition that men and women scientists receive some five to eight years after their doctorates, and we have concluded that there is little discrimination against women scientists after receipt of the Ph.D. It is possible, however, that a social system can act both universalistically and rationally toward a group of scientists at one point in time and still be discriminating against that group. If the conditions for a self-fulfilling prophecy are set up at time 1 then, of course, a rational and universalistic judgement at time 2 will ultimately produce inferior status for the judged group.[40] More specifically, let us suppose that the social system of science is less likely to support the education of women with graduate fellowships and is less likely, once she has received her degree, to give her financial support for the resources and facilities necessary to carry out research. If women scientists were to receive less support, then it would not be surprising if they turn out to be less productive than men. Their futures then become predictable. When they come up for hiring and promotion decisions, their publication records are carefully reviewed, found inferior to those of men with the same type of background, and they lose out in the academic marketplace. At this point the self-fulfilling prophecy based on the assumption that women are simply less motivated and less productive than men would appear to be strikingly supported. Gatekeepers for resources would feel justified in not giving financial support to women scientists since they are less likely to produce significant research with the funds. And so the conditions for the self-fulfilling prophecy would be reinforced.

Critics of the reward system of science hypothesize that women scientists suffer from the processes of accumulating disadvantage. Once denied access to resources and facilities, they are forever more involved in a struggle simply to reach parity with their male

39. For another discussion of how these processes affect the careers of scientific elites, see Zuckerman, *Scientific Elites.*

40. For a full discussion of the structural conditions making for self-fulfilling prophecies, see Merton, *Social Theory,* pp. 475–90.

colleagues. Harriet Zuckerman has recently referred to this process as the "principle of the double penalty," a principle which holds that groups like blacks and women not only suffer from direct discrimination but also from the second penalty of being placed initially into second-rate structural positions which make it almost impossible for them to produce the outstanding work that is necessary for moving out of such positions.[41]

Before accepting this inviting theoretical explanation for the lower productivity and less distinguished positions held by some female scientists, we must ask whether it is empirically correct. Are women less likely to be admitted to top graduate departments of science, independently of their ability? Do women receive less financial support when in graduate school than men? Are women's efforts at research given less support during their training period than men? Are women working on research projects given the less intellectually demanding jobs to perform? Must women more than men choose dissertation topics which are less compatible with their interests? We cannot answer all of these questions at this time. More research is needed on issues which involve the initial definition of talent and potential of men and women scientists.

Some evidence is available on the extent to which women and men are admitted to and given financial support for their graduate training. Although the evidence is scanty, women appear to receive their doctorates from top-ranked departments in the same proportion as men. Folger, Astin, and Bayer examined the proportion of men and women Ph.D.'s from departments rated as distinguished or strong in Cartter's ACE study of the quality of graduate education.[42] They found that about 50 percent of each sex were being trained at these top departments. Berelson, in an earlier study of graduate education, also found that women were just as likely as men to receive degrees from one of his "top twelve" schools.[43] Data reported by Astin on admission to medical schools show similar results: in 1964–65, "47.6 per cent of the women applicants were accepted, as compared with 47.1 per cent of the male applicants."[44]

If we turn to the distribution of fellowship support for men and women graduate students, we find no prima facie evidence of discriminatory practices. Women are just as likely as men to receive

41. Zuckerman, "Women and Blacks."

42. John K. Folger, Helen S. Astin, Alan E. Bayer, *Human Resources and Higher Education* (New York: Russell Sage, 1970).

43. Cited in *Academic Women,* p. 89.

44. Astin, *The Woman Doctorate,* p. 103.

financial support. Astin, comparing the proportion of women and men doctorates receiving either fellowships or graduate assistantships, found that 57 percent of the female doctorates had received aid compared with 58 percent of the men.[45] Jessie Bernard also reported: "The National Science Foundation awards in 1959 were given to women in about the same ratio as to men; 12 percent of the applicants were women, and 12 percent of the awards went to women."[46] She notes that women receive fellowships to the Center for Advanced Study in the Behavioral Sciences at Palo Alto in proportion to the numbers of females recommended for them. She concludes: "However convincing individual cases of prejudiced discrimination are, it is difficult to prove its existence on a large or mass scale. The most talented women may be and, indeed, are victimized by it, but apparently not academic women en masse. At least the evidence from awards and from the number of academic women in proportion to the qualified pool available is far from convincing."[47] To add only three more pieces of evidence to those already presented, consider the conclusion reached by James A. Davis in his study of some 34,000 college graduates in 1961. He concluded that "women . . . have no disadvantage or advantage in offerings" of fellowships for graduate education.[48] Simon, as noted above, reports that women are slightly more likely than men to receive some form of fellowship support. And finally, our own data showed no significant differences between the number of postdoctoral honorific awards received by men and women in the matched sample.

Clearly, these data only suggest that women are not treated differently than men in terms of support necessary to do full-time graduate work. It is possible, of course, that the average woman who applies for admission to graduate school and for financial assistance is significantly more able than the average male applicant. If this were demonstrated, we would expect that women would be both admitted and awarded fellowships at a disproportionately high rate. Available data from Graduate Record Examination scores of men and women applicants for graduate admission do not support this hypothesis. Men do better than women on the quantitative part of the test, and women do slightly better than men on the verbal

45. Ibid.
46. Bernard, *Academic Women,* p. 50.
47. Ibid.
48. Quoted in ibid., p. 51.

part.[49] We would conclude that the scanty data which do exist offer little support for the hypothesis that women are the victims of a self-fulfilling prophecy and suffer from accumulative disadvantage. Of course, a great deal more research is needed before any final conclusions can be drawn.

The primary aim of this section was to examine a set of common beliefs about the treatment that women scientists receive in the academic science community. Proving discrimination is no easy business. Apparent discrimination often turns out to be illusory, a consequence of the interaction of several factors with the variable which is perceived as being the basis of discrimination. For example, we have suggested that studies of recognition- or status-attainment in science which do not adequately take into account variations in the role-performance of the persons involved can have little value in establishing actual as distinguished from apparent discrimination. Fortunately, we have been able to place some controls on both the initial talent and the research performance of the scientists in this sample.

What conclusions can we draw from these data? Some differential treatment of women does obtain in the social system of science, but not as much as many critics of the scientific reward system would have us believe. Women scientists are every bit as natively able as male peers who receive their training at equivalent departments. Yet women scientists, regardless of their marital status or the size of their families, simply produce fewer scientific papers, and papers of less impact, than men in comparable positions. When quality and quantity of publications are taken into account, sex status has only a minor effect on the prestige of the scientist's academic affiliation. Sex status does have a significant independent effect on the overall academic rank of scientists. Women are not as frequently promoted to senior positions especially at the better universities, as are the men who come from the same doctoral departments.

In light of the data on the treatment of women scientists, is it necessary to alter our conclusions about the extent to which the

49. In *GRE: Guide to the Use of GRE Scores in Graduate Admissions, 1971–72* (Princeton, New Jersey: Educational Testing Service, 1971), p. 19. For the period 1968–71, the mean verbal score for men was 499; for women, 511. The mean quantitative score for men was 549; for women, 472. A more reliable comparison would, of course, show the comparative scores of only those men and women applying to graduate schools in scientific fields. These data are not currently available in a published source.

scientific stratification system is universalistic. Again, we must make clear that we are only considering the treatment of women scientists after the receipt of the Ph.D. The first and most important fact is that sex status explains very little variance on recognition of all types. The zero-order correlations vary between zero and .30. Thus, even if discrimination were occurring, it would not explain much variance on rank in the stratification system.

The zero-order correlations between sex status and recognition seem to result from three basic factors. The first is self-selection. Women may to some extent be in less prestigious institutions than men because they have less mobility. Many women scientists are limited in where they can get jobs by the occupational contingencies of their husbands. The fact that a woman's job has in the past been defined as "less important" than a man's is certainly a type of discrimination; but it is not discrimination within science. If a married woman turns down a job at a university in a locale removed from her husband's place of employment, we cannot say this is evidence of discrimination in science.

The second reason for the small correlation between sex status and recognition is that women produce fewer and less frequently cited papers than men. All the traditional explanations for these productivity differentials remain as unsupported hypotheses. None of the data that we have, nor any of the literature on women scientists that we know wash out these differentials. They are definitely not a result of family status or of differences in institutional location. Unmarried women produce less than men. Women at universities produce less than their male colleagues. Why then do women produce less than men? Data on I.Q.'s suggest that men and women have at least equal native ability. Although this must remain as speculation, we believe that productivity differences will ultimately be explained by differences in motivation. Due to socialization and the value system in the larger society, it is possible that in the past, women have not been as committed to their careers and not as driven to achieve the heights of success.[50]

It is possible that receiving the doctorate has a qualitatively dif-

50. A number of recent studies have discussed the impact of socialization on the achievement orientation of women. See Ralph Turner, "Some Aspects of Women's Ambition," *American Journal of Sociology* 72 (September 1966):163–72; Alice S. Rossi, "Why So Few?"; Patricia A. Graham, "Women in Academe," *Science* 169 (25 September 1970):1284–90; C. Bird and S. W. Briller, *Born Female: The High Cost of Keeping Women Down* (McKay, New York, 1968).

ferent meaning for men and women. For men, the doctorate represents the "union card" necessary for entry into the profession, but it is in no way sufficient to achieve success in science. For women, the act of obtaining the Ph.D. with the attendant struggles that the process entails, places them in elite positions relative to their sex peers. These difficulties are indicated by the higher proportion of female than male doctoral students who drop out of graduate programs.[51] Further, women tend to take longer to finish their degree. To have received the degree may be viewed and indeed, may be for the woman, the end of an odyssey. To have earned the degree is, in some measure, a triumph. The motivation that is required to set out full-steam ahead upon the production of scientific research may not reach the same level of intensity for both men and women doctorate recipients. And if women fail to be as productive in the years immediately following their degree, the social process of accumulative disadvantage may take over and contribute to their falling further behind in the race to produce new scientific discoveries. In short, the productivity of the female scholar may depend somewhat on her reference group and her motivation. If the comparative reference group of women scientists is other women scientists or women in the larger societal context rather than the entire scientific community, women may be less motivated and therefore less productive than men. We hypothesize that using a satisfactory measure of motivation, sex differentials in productivity will be reduced.

The third and final factor creating the small correlation between sex status and recognition is discrimination. In the past, there probably was some discrimination against women in science and there may still be some small amount of discrimination. In our data, discrimination is most evident when considering the effects of sex on academic rank. In 1965 it probably took women slightly longer than men to be promoted from assistant to associate professor. Whether this is the situation today would require further research. In general, we conclude that the amount of discrimination against women scientists is small—certainly less than the "discrimination" faced by graduates of low-prestige academic departments. The data on sex

51. Allan Tucker, David Gottlieb and John Pease, "Attrition of Graduate Students at the Ph.D. Level in the Traditional Arts and Sciences" (Office of Research Development and the Graduate School, Michigan State University, East Lansing, Publication No. 8, 1964). In a study of twenty-four universities selected on the basis of size, region, and quality of their graduate program, this study found that "among female doctoral students 54 percent dropped out, whereas among male doctoral students, the attrition rate was 36 percent" (p. 57).

do not require a modification of our conclusion that the scientific stratification system is basically universalistic and rational.

Thus far we have explored only the possibility that one functionally irrelevant status influences status attainment in science. There are, of course, other social characteristics of individual scientists which could influence their level of recognition, regardless of the quality of their role performance. We shall consider two of these: racial and religious status.

RACIAL STATUS AND SCIENTIFIC RECOGNITION

Of the various bases of discrimination in American society, none is more salient than racial status. But does science treat racial minorities, and in particular black Americans, fairly? We may feel sure of only one point in addressing this question: there are almost no detailed, reliable data on the position of blacks in American science. The scattered empirical evidence that does exist all converges upon a singular fact: there are few blacks who have Ph.D.'s and virtually no blacks in contemporary American science. Since receipt of the doctorate is a prerequisite for entry into the upper levels of academic science, we first consider the proportions and absolute numbers of doctorates received by blacks.

Crossland surveyed sixty-three Ph.D.-granting institutions to compile data on the distribution of blacks in graduate departments. The picture painted is bleak: "1.72 percent . . . of the total enrollment in America's graduate schools of arts and sciences" are black Americans, and "0.78 percent . . . of all Ph.D.'s awarded between 1964 and 1968 went to black Americans."[52] While the absolute number of blacks seeking and obtaining doctorates has sharply increased since World War II, the percentage of the total number of new doctorates that blacks represent has remained fundamentally unchanged.[53]

There are various estimates of the actual number of blacks within the scientific community; all agree that blacks receive less than 1 percent of science Ph.D.'s conferred in any given year. A recent survey indicates that black Ph.D.'s are not evenly distributed among the various fields of higher education. Twenty-nine percent of black Ph.D.'s have received their doctorates in education; 26 percent in

52. Fred E. Crossland, "Graduate Education and Black Americans" (New York: Ford Foundation, 1968, mimeo), p. 2.

53. James M. Jay, *Negroes in Science: Natural Science Doctorates, 1876–1969,* (Detroit: Balamp Publishing, 1971), p. 2; Crossland, "Black Americans," p. 3.

the social sciences. Approximately 25 percent of black doctorates receive their degrees in one of the physical or biological sciences.[54] Jay estimates that a total of "650 American Negroes obtained doctorate degrees in the natural sciences between 1876 [when the first doctorate to be granted to a black American was conferred at Yale] and 1969."[55]

Crossland could locate only one black doctorate in physics, eight in chemistry, and eleven in biology, among all recipients of the Ph.D. at sixty-three universities during the two-year period of 1967 and 1968. Consider what a minute fraction this represents when in the same years there were 3,546 doctorates awarded in chemistry, 2,712 in physics and astronomy, and 6,797 in biological sciences.[56] Jay's research could identify only 94 black Ph.D.'s in the physical sciences and 252 doctorates in the biological sciences between 1876 and 1969.[57] Even if these estimates are off by a fairly wide margin, it should be clear that, if there is a paucity of women doctorates in science, there is even a greater scarcity of black Americans in science.

Estimating the treatment of blacks in science is, therefore, virtually impossible. Even if we deal with an entire population rather than samples of scientists, we quickly begin to consider individual cases. Furthermore, there have been to date no studies on the role-performance and productivity of black scientists. What we know is limited indeed: black doctorates tend to receive their degrees from less distinguished universities; they take longer, on average, to earn their degrees; once they have the doctorate, they are most likely to be located at black colleges far removed from the frontiers of scientific advance.[58]

We must, of course, ask why there are so few blacks in science. We have no firm answers to this question. We do know that blacks, like women, turn away from the sciences at comparatively early ages. Black children, when viewing occupations, see science as open to only the most exceptionally able students. In fact, they feel that science is a more difficult career to pursue than sixteen other

54. Ford Foundation, *A Survey of Black American Doctorates* (New York: Ford Foundation, 1970).

55. Jay, *Negroes in Science*, p. vii.

56. *Summary Report, 1967 Doctorate Recipients from United States Universities; Summary Report 1968 Doctorate Recipients from United States Universities.*

57. Jay, "Negroes in Science," pp. 40, 50.

58. Crossland, "Black Americans"; Ford Foundation, *Black Doctorates;* Jay, *Negroes in Science.*

professional and managerial occupations.[59] Further, Harmon has shown that scientists tend to come from the homes of well-educated, professional parents.[60] In this respect, black Americans are clearly at a disadvantage, being far underrepresented among the group of middle-class professionals.

It would appear, then, that black Americans face many hurdles along the road to science. Not only must they overcome the disadvantages of their poorer socioeconomic backgrounds, but they must surmount the generally poorer training they have received at less distinguished undergraduate and graduate schools. Having been trained at locations not oriented toward the production of new scientific research, and which have few facilities to carry out research ideas, blacks face the prospect of being penalized twice: "once by the direct psychic costs of having been discriminated against" prior to entrance to colleges, "and again by the subsequent restricted access to resources" once they become scientists.[61] Although we cannot at this time assess whether there is any actual discrimination against black Americans *within* the scientific community in just access to scientific resources and facilities and in recognition accorded for outstanding role-performance, it is clear that prior to entrance into the scientific community, even the most talented blacks face a set of social and psychological barriers that are difficult to overcome. In short, since we have no data on the ability or the scientific output of black scientists, it is currently impossible to say with any certainty whether the positions that blacks occupy in science are the result of a self-fulfilling prophecy.

RELIGIOUS STATUS AND SCIENTIFIC RECOGNITION

Thus far we have considered two statuses exogenous to the scientific social system: sex status and race. Although other statuses such as age may prove to have an independent effect on the evaluation of scientists, we have collected data on only one other nonprofessional status which is sometimes said to be an influence on the status attainment of men and women in academic life: religion. For the pur-

59. Joseph H. Fichter, *Graduates of Predominantly Negro Colleges: Class of 1964* (Washington, D.C.: U.S. Government Printing Office, Public Health Service Publication #1571, no date). These findings are described in Zuckerman, "Women and Blacks."

60. Lindsey R. Harmon, *Profiles of Ph.D.'s in the Sciences* (Washington, D.C.: National Academy of Sciences, National Research Council, Publication #1293, 1965).

61. Zuckerman, "Women and Blacks," p. 36.

poses of this study we limit our inquiry to an analysis of the effect of being Jewish on location in the stratification system. Many scholars have expressed the view that anti-Semitism has been the major form of religious discrimination operating in academic social systems over the past century.[62]

Our starting point for the analysis of the influence of sex status on recognition in science was to note the underrepresentation of women in scientific occupations. When we examine the representation of Jews in science, the opposite pattern obtains. Weyl and Possony, drawing upon George Sarton's work on the history of science, note that in the first fourteen hundred years of Christendom, Jews were heavily represented among scholars. Some 10.6 percent of the 1,897 scholars listed by Sarton were Jewish, more than three times their proportion in the European population.[63]

In the recent past, Jewish scholars and scientists have been well represented among recipients of prestigious awards. From the inception of the Nobel Prize to 1965, some 27 percent of American Nobelists in science were Jewish; throughout the world, Jews accounted for 16 percent of the 225 science laureates between 1901 and 1962.[64]

The intellectual contribution of Jews has been underscored by many writers on the history of culture. In 1919, Thorstein Veblen remarked:

> It is a fact which must strike any dispassionate observer that the Jewish people have contributed much more than an even share to the intellectual life of modern Europe. So also is it plain that the civilization of Christendom continues today to draw heavily on the Jews for men devoted to science and scholarly pursuits. It is not only that men of Jewish extraction continue to supply more than proportionate quota to the rank and file engaged in scientific and scholarly work, but a disproportionate number of

62. For discussions of anti-Semitism in universities and colleges, see among others, Heywood Broun and George Britt, *Christians Only* (New York: The Vanguard Press, 1931); Carey McWilliams, *A Mask for Privilege: Anti-Semitism in the United States* (Boston: Little, Brown, 1948), pp. 38–39; Don W. Dodson, "College Quotas and American Democracy," *The American Scholar* (Summer 1945); Lawrence Bloomgarden, "Our Changing Elite Colleges," *Commentary,* February 1960; Seymour Martin Lipset and Everett Carll Ladd, Jr., "Jewish Academics in the United States: Their Achievements, Culture and Politics," *American Jewish Yearbook* (New York: The Jewish Publication Society of America, 1971):89–128.

63. Nathaniel Weyl and Stefan Possony, *The Geography of Intellect* (Chicago: Henry Regnery, 1963):123–28.

64. These data were reported in Lipset and Ladd, "Jewish Academics," p. 97.

the men to whom modern science and scholarship look for guidance and leadership are of the same derivation. . . . They count particularly among the vanguard, the pioneers, the uneasy guild of pathfinders and iconoclasts, in science, scholarship, and institutional change and growth.[65]

All of this testimony to the preeminent position of Jews in the history of scholarly and scientific pursuits actually says little about whether Jews have been the objects of systematic discriminatory practices. One can make a long list of notable Jews who have "made it" in academic and scholarly life, show that they are overrepresented in relation to their proportion in the general population, and still conclude that there is discrimination against Jews. As we have noted above, representation relative to comparative population statistics tells us almost nothing about equality of opportunity. For who is to say, without systematic review, that a far higher proportion of Nobelists, or leading architects of the house of intellect, would have been of Jewish origins had not anti-Semitism existed in the academic and scientific social systems.

There is no way that we can test, at this time, the presence or the extent of discrimination against Jewish scientists in the past. This we must leave to the historian of science. We can begin to examine more closely, however, whether Jewish scientists are fairly treated by the scientific community today. In order to test whether the functionally irrelevant status of religion is activated in specific structural situations, we must estimate the independent influence of religious status on forms of recognition.

A sample of 300 full professors in five disciplines at Ph.D.-granting institutions was used to determine the influence of religious status on recognition.[66] We did not have data on self-reported religious preference. In order to determine the religion of the scientists, we had one Jewish faculty member and two Jewish graduate students independently guess the religion of the scientists from their names. Where two of the three judges agreed that a name was Jewish, the scientist was so classified.[67] Of course, using this procedure, we are

65. Thorstein Veblen, "The Intellectual Pre-Eminence of Jews in Modern Europe," *Essays in Our Changing Order* (New York: The Viking Press, 1934), pp. 221, 223–24, as quoted in Lipset and Ladd, "Jewish Academics," pp. 96–97.

66. For a description of this sample of 300 full professors, see Appendix A.

67. For a discussion of a similar procedure in identifying Jewish professionals, see Stephen Cole, *The Unionization of Teachers* (New York: Praeger Publishers,

bound to misclassify some of the scientists. But since the three guessers knew nothing about the scientists, there is no reason to believe that the errors would not be randomly distributed. Using this rough procedure, we classified 68 out of the 300, or 23 percent, as Jewish. This figure is fairly consistent with independent estimates of the proportion of Jews in academic science. Lipset and Ladd, in their study of some 60,000 academics at colleges and universities throughout the United States, found that approximately 14 percent of the physicists, 8 percent of the chemists, 21 percent of the biochemists, 13 percent of the sociologists, and 17 percent of the psychologists, had Jewish social origins.[68] They also pointed out, however, that Jews are more likely than non-Jews to be located at universities. Since all our scientists are at Ph.D.-granting institutions, the figure of 23 percent is at least roughly consistent with the data reported by Lipset and Ladd.

In the analysis, we use three dependent variables: one form of positional success, indicated by the prestige rank of a scientist's academic affiliation; one form of reputational success, measured by a scientist's visibility to the community; and the perceived quality of his work.[69] The zero-order correlations between religion and the three dependent variables are significantly greater than zero, with Jews receiving more recognition than non-Jews. Jews are also more likely to have their work perceived as being of high quality. The correlation between religion and rank of department is $r = .16$; between religion and visibility, $r = .27$; religion and perceived quality, $r = .19$. These coefficients suggest that Jewish scientists are slightly more likely than non-Jews to be recognized in contemporary science. However, just as proportional representation of a group in science and in the general population tells us little about whether discrimination is at work, neither do zero-order correlations. It is possible that, even though Jews are more likely to be recognized, they still may be discriminated against. If twice as many Jews as non-Jews have produced high-quality scientific work, then twice as many

1969), pp. 94–95. The procedure was first introduced by Paul F. Lazarsfeld. He had people guess the religion of academics from their names. Since he also had data on the actual religious preference of the academics, he was able to estimate the accuracy of this guessing procedure. He found this guessing procedure to yield roughly accurate estimates of the proportion of Jewish and non-Jewish academics.

69. For the exact question used to measure "perceived quality," see chapter 4, p. 105.

Jews as non-Jews should be eminent. If Jews are only slightly more likely to be eminent, then discrimination could be operating.[70] We must therefore look at the quality of role-performance of Jewish and non-Jewish scientists.

Lipset and Ladd report that 21 percent of Jewish scholars included in the Carnegie sample indicate that they have published more than twenty articles in academic or professional journals, compared to only 9 percent of the Protestant and 6 percent of the Catholic aca-

Table 5. The influence of religious status on two forms of scientific recognition and perceived quality of work, controlling for scientific output

	Zero-order correlations and regression coefficients in standard form					
	Dependent variables					
Independent variable	Rank of dept.		Visibility		Perceived quality	
	r	b*	r	b*	r	b*
Quantity†	.21	.00	.49	.23	.44	.14
Quality of Work‡	.35	.34	.59	.43	.58	.48
Religion§	.16	.14	.27	.24	.19	.15

†Productivity was measured by counting the total number of published papers listed in the journal abstracts of the five scientific fields used for this study.
‡Quality was measured by counting the total number of citations to the scientists' lifework in the 1965 and 1969 *SCI*.
§Religion was coded: 2 = Jews, 1 = all others.

demics. Only 26 percent of the Jewish academics report not having published any articles, while 51 percent of the Catholics and 44 percent of the Protestants report no publications.[71] In our sample of 300 full professors, however, we find virtually no correlation between religious status and the quantity or quality of research output ($r = .00$ and $r = .07$, respectively). There are at least two possible explanations for the difference between the results from our data

70. In order for discrimination to be statistically demonstrated, when we control for quality of output, the standardized regression coefficient between religion and recognition would have to be negative. In the statistical sense this same negative coefficient would have to obtain regardless of whether the independent variable being analyzed was religion or sex status.

71. Lipset and Ladd, "Jewish Academics," p. 101.

and those of Lipset and Ladd. Their finding is based upon self-reported productivity data. It is possible, although unlikely, that Jews would be more likely than non-Jews to exaggerate their publication records. More probably, the difference can be explained by the fact that all the scientists in our sample were located at Ph.D.-granting institutions. We would guess that if Lipset and Ladd were to compare the productivity of Jews and non-Jews only within Ph.D.-granting institutions, they would find little if any difference.

Since, in our data, there is no relationship between religion and scientific output, and since Jews are slightly more likely to be recognized than non-Jews, we may conclude that there is no evidence that Jewish scientists are discriminated against. As the data of table 5 indicate, religious status does have a slight independent effect on recognition, but an effect which in no way compares to that of quality of research. We may tentatively conclude that there is no behavioral discrimination against Jewish scientists as a group in the American academic science community. It is of course possible that we might be able to identify individual cases of anti-Semitism. Our data deal only with rates and group patterns.

Conclusion

The general conclusions to be drawn from the data presented in this chapter are clear: there appears to be little empirical support for the claims of those who hold that science actively discriminates against women and Jewish members of the scientific community. Since there are virtually no black scientists in the United States, there is little opportunity, even if the predisposition were present, for the scientific community to consciously deny black scientists high social positions.

There is obviously a great deal of discrimination against women and blacks in the larger society. How do we account for the seeming absence of discrimination in science? We believe that if the educational establishment is considered as a pyramid, with elementary schools on the bottom and elite scientific institutions at the tip, the higher up the pyramid one goes the less discrimination will be encountered. Blacks will face more discriminatory conditions in elementary school than in high school, more in high school than in college, more in college than in graduate school, and probably more in graduate school than in the profession. We mean by discriminatory conditions overt forms of discrimination and subtle forms of discrimination, such as guidance counsellors urging blacks to pursue a vocational course in high school or urging women to major in the humanities rather than the sciences.

Such a pyramid of discriminatory conditions could possibly result from two factors. First, the higher one goes in education, the clearer and more rational are the criteria upon which people are evaluated. Thus, although well-behaved elementary school pupils get higher rewards than poorly-behaved students, "conduct" is rarely used as a criteria of evaluation in college. Also, it is easier to tell whether someone has published a useful scientific paper than whether they should get an A or a B in a college course. Second, we would guess that the higher up the educational ladder one climbs, the more status-judges and gatekeepers are committed to the application of universalistic criteria of evaluation. In other words, blacks and women are far more likely to meet discriminatory conditions on the way to becoming a scientist than they are in the profession itself. However, blacks and women who have survived the ordeals of the journey may still feel the effects of past discrimination after they enter science. For example, the lower productivity of women scientists may result from past trials and tribulations rather than any active discrimination in science itself. These are, of course, speculations and would require a great deal of further research to substantiate.

6 STRATIFICATION AND THE COMMUNICATION OF SCIENTIFIC INFORMATION

The two basic problems we examine in this book are the processes through which scientists find their level in the stratification system and the consequences for the scientists of being located at a particular level. In this chapter we present data bearing on the second of these problems. We will be particularly interested in how rank in the stratification system influences receipt of information. In science, communication is basically open. The value system stresses that knowledge is public; secrecy is prohibited. Most scientists publish their discoveries in journals which are available in libraries to anyone who chooses to use them. Scientific meetings are open to all interested people. Most scientists will freely send preprints of their articles to any colleague requesting one. So called "invisible colleges" are merely groups of scientists working in the same specialty who exchange preprints and occasionally meet at conferences.[1] Invisible colleges themselves are relatively open. Usually, any scientist working in the specialty can get on the mailing lists to receive preprints.

There are, of course, some forms of research which are kept secret. Military research for the Department of Defense and some industrial research often involve closed communication. However, most of this research is applied rather than basic, and in this book we are concentrating on basic research. Results of research are often kept secret by scientists while they check their data for possible errors. Occasionally, scientists who fear being anticipated will withhold information about their research in the period after the work is completed but before it is submitted for publication. This period is a relatively short one: on average, four months in the physical sciences; seven months in the social sciences.[2] In short, it would

1. For discussion of invisible colleges, see Price, *Little Science, Big Science;* Crane, *Invisible Colleges.*

2. See William D. Garvey, Nan Lin, and Carnot E. Nelson, "Some Comparisons of Communication Activities in the Physical and Social Sciences," in C. E. Nelson

be rare to find cases where information is willfully withheld from other scientists for more than a short period of time, since the withholding of information is considered to be a form of deviant behavior. Thus, when we study communication in science, we are studying a system which is at least formally open to all. Differences in levels of knowledge possessed by scientists will result more from how the scientist ties himself into the system than from formal structural barriers to communication.[3]

Scientific communication involves the exchange of information between the producers of discoveries and an audience consisting of other scientists working in the field. In chapter 4 we analyzed those characteristics of scientists and their work which influence their visibility. Since communication involves both authors and audience, to fully understand how social stratification affects communication, we must examine how characteristics of the audience influence the flow of information. Just as there are conditions influencing the visibility of a man's work, there may also be conditions influencing the extent to which a scientist is tied into the communication network. Just as some men can be easily seen (that is, have high visibility), others are in positions where they can easily see. This latter characteristic we call "awareness." The difference between visibility and awareness is the difference between passive and active. Visibility characterizes the men being looked at; awareness, the men who are looking.[4]

Awareness is basically analogous to the concept of "effective scope."[5] We can view the scientific community as a city, with a variety of neighborhoods and many social and economic strata. In

---

and D. K. Pollock, eds., *Communication Among Scientists and Engineers* (Lexington, Mass.: D. C. Heath and Company, 1970), p. 62.

3. There are, of course, purely informal types of communications that often play a significant part in the evolution of discoveries. There is also private correspondence which is exchanged between scientists. Here we shall concentrate on the communication of published ideas.

4. Visibility and awareness have been shown to be concepts relevant for the analysis of many types of social activity, not, of course, only for the study of communication systems in science. For a discussion of other types of human behavior for which these concepts are relevant, see Merton, *Social Theory,* chapter 9. In that chapter, Merton uses the terms "visibility" and "observability." We substitute the term "awareness" for "observability."

5. The concept of "effective scope" was first introduced by Paul F. Lazarsfeld in his early study of Austrian working-class youth. It was further developed in Elihu Katz and Paul F. Lazarsfeld, *Personal Influence* (Glencoe, Illinois: The Free Press, 1955); Paul F. Lazarsfeld and Wagner Thielens, Jr., *The Academic Mind* (Glencoe, Illinois: The Free Press, 1958).

this city there are public parks, museums, and other public facilities located in different sections, but most of them are in a "downtown" area. These facilities are open to everyone; they may be used if the individual wishes to make use of them. Yet some individuals make far greater use of these facilities and cultural resources than others. They know about more of the facilities, that is they have a broader "effective scope," and they utilize the available facilities more effectively. Various studies have shown that social class influences the degree of knowledge and use that people make of various types of facilities. For instance, Caplovitz's study of the purchasing patterns of the poor suggested that poor people have narrower "effective scopes" and are not as likely to leave their neighborhood to make use of available resources in other sections of the city.[6] Communication in science may be somewhat analogous to this situation. Although any scientist can, if he wishes, make use of the communication facilities, not all do. Can we say that observed differences in the "effective scope" of scientists result from their location in the stratification system?

AWARENESS OF SCIENTIFIC RESEARCH

Not all scientists have an equally broad knowledge of work done in their field. This introduces the question of whether these differences in knowledge are due to idiosyncratic characteristics or to socially structured differences in awareness. The question we address here is the extent to which a scientist's awareness is influenced by location in the stratification system. Do physicists at top-ranking departments have more knowledge of what is going on in their field than those at the lesser departments? Are men working at eastern universities as thoroughly familiar with work being done at western universities as with work done closer to home?[7] Do physicists producing high-quality work have a greater awareness than those doing less significant work? To study awareness, we used the sample of 1,308 university physicists. Awareness was measured by the same question that yielded visibility scores:

> These twenty-five physicists are drawn from various universities, institutes, and fields of investigation. You may not know the work of some of these men, but please indicate, for each case,

6. David Caplovitz, *The Poor Pay More* (New York: Free Press of Glencoe, 1963).

7. Although region is not a stratification variable, it is significant in analyzing the flow of communications.

the extent of your familiarity with their work by circling the appropriate number: (1) familiar with most of his work, (2) familiar with some of his work, (3) familiar with a small part of his work, (4) unfamiliar with his work but have heard of him, (5) have never heard of him.

If a respondent gave either of the first three answers, he was considered to be aware of the work of the physicist. We did not differentiate between the first three answers since we thought that scientists might use different frames of reference in estimating how much of a man's work they were familiar with.

Before beginning our discussion of the determinants of awareness, we must make explicit an underlying assumption in this analysis. Our measure of "awareness" is the extent to which scientists report that they are familiar with the work of twenty-four physicists. (One of the twenty-five names in the questionnaire was fictitious, used to estimate exaggeration of knowledge.) When a scientist reports that he is familiar with another's work, what does this mean? We assume that such a report means more than simple familiarity with the scientist's name; category 4 indicated this type of knowledge. At the same time, we do not assume that awareness of a scientist's work means possession of detailed knowledge of that work. The type of knowledge indicated by awareness is that which could at least enable the respondent to give a brief summary of the type of work that the physicist has done in the past—or, in some proportion of cases, give a far more detailed account of the work.[8] Thus, when a solid-state physicist says he is familiar with the work of the elementary-particle theorist, T. D. Lee, this is assumed to mean that he could give a brief description of Lee's most significant contribution. Such knowledge is important as it indicates a broad familiarity with problems and perspectives in one's field. There is no way, of course, that we can be sure that answers to the question actually indicate possession of this type of knowledge. We must take this as an assumption until more precise research can be completed.

A physicist's awareness score is the number of men whose work he knew about.[9] The distribution of awareness scores is presented in table 1. Awareness varies less than visibility; the coefficient of

8. There was, of course, variability in the meaning that different respondents gave to the question. Two physicists say they are very familiar with the work of another physicist. The first may have a detailed understanding and appreciation of his work; the second only superficial knowledge.

9. For a discussion of the validity of this measure of awareness, see note 15 below.

variation for visibility was .68, compared with .35 for awareness. The majority of physicists knew the work of about half the men in their questionnaire. The relatively small amount of variance on awareness was the first sign that we would find communication to flow evenly to all corners of the system and that we were unlikely to discover great differences in socially structured awareness of scientific research.

Table 1. Awareness scores

Number of men known		Number of observers	
0–5		122	(9%)
6–8		268	(21%)
9–12		531	(41%)
13–15		265	(20%)
16–24		122	(9%)
	Total	1,308	(100%)

Mean = 10.5
Standard Deviation = 3.73
Coefficient of Variation = 35%

The hypothesis that there would be few socially structured differences in awareness was confirmed by a multiple regression analysis. All of the independent variables for which we had data could explain a total of only 19 percent of the variance on awareness.[10] In table 2 we present the correlation coefficients between awareness and several aspects of a scientist's location in the stratification system that we thought might influence awareness. These correlation coefficients indicate that the structured differences in awareness are small.

Age and professorial rank are most highly correlated with awareness. These two variables are, of course, related to each other ($r = .53$). Only 7 percent of the physicists who were fifty years old or older were not full professors. Age has little independent effect on awareness, yielding a partial correlation of .09 when professorial rank is held constant. Professorial rank and awareness, however, have a partial correlation of .20 when age is held constant. The data indicate that it is not so much the accumulation of years of

10. The fact that we were unable to explain a large amount of variance on awareness could mean that we did not employ the relevant sociological variables or that the small amount of variance on this variable is the result of idiosyncratic factors.

experience that makes for high awareness as the reaching of "professional maturity"—the achievement of high academic rank. Full professors are more likely to supervise large numbers of doctoral dissertations, review applications for grants, and be on the editorial boards of journals.[11] These activities probably contribute to an awareness of scientific research.

Perhaps most significant is the fact that both rank and age together explain very little (8 percent) of the variance on awareness. Even the two variables most highly correlated with awareness make for little differentiation, indicating that in physics, information is assimilated at an equal rate at all levels of the stratification system. A physicist's eminence (as measured by awards) and the quality of

Table 2. Coefficients of correlation between awareness and its potential producers

Independent variable	Zero-order correlation with awareness
Professorial rank	.28
Age	.22
Prestige of highest award	.19
Number of awards	.18
Rank of current department	.17
Quality of work (number of citations)	.15

N = 1,308.

his work (as measured by citations) also explained little variance on awareness.

The variable we were most interested in was rank of department. We thought that physicists who taught at top-ranked departments would be closer to the center of the informal communication system of physics and thus have greater awareness. The data show a difference in the expected direction, but a small one ($r = .17$). Men at the top departments had slightly greater awareness of research in physics than their colleagues at lower-ranking departments. Is this weak but nonetheless positive correlation a result of location in different departments, or can it be explained by the differing personal

11. It would, of course, be useful to have these data. If we knew which physicists actually did supervise a large number of doctoral dissertations, reviewed applications for grants, and were on the editorial boards of journals, we could then empirically test the validity of the interpretation given here.

characteristics of scientists at departments of varying prestige? The data indicate that rank of department has an independent effect on awareness. There is virtually no reduction in the correlation between these two variables when the number of awards held by a scientist is controlled (partial r = .15). Similarly, the data indicate that knowing the quality of work done by men in different departments does not significantly reduce the correlation between department rank and awareness (partial r = .14). However, this correlation procedure covers up a specification of the relationship that tabular data reveal. When we computed the mean number of men identified by physicists at the different departments and then controlled for the number of awards they held, we found that rank of department made no difference in awareness among those who had more than one award.[12]

Although there are only minor differences in awareness, it is possible that these differences will increase for specific characteristics of the objects being observed. Until now, we have been considering what might be called "gross awareness"—awareness of the system as a whole. In the next part of the analysis, we shall vary characteristics of the observed scientist and observers simultaneously.

We begin with an analysis of geographical location. Although region is not a stratification variable, it is important in assessing the openness of communication. If ideas flow from research centers in one part of the country to research centers in all other parts, people in the East will be just as familiar with work done in the West as will people who work in the West. As can be seen in table 3, this is approximated in the case of physics. Region makes for no systematic differences in awareness. (This may be seen by looking at the rows of table 3.) Men working in the East were no more visible to eastern colleagues than they were to those working in other regions. We thought it possible, however, that if we further divided the physicists being observed into those who had produced high-quality work and those who had not, we might find some differences. It might be that high-quality work would be clearly visible in all parts of the country regardless of where it was produced; but less frequently cited work might be more visible to the men who worked in the region where

12. The correlation analysis did not reveal this because there was, of course, a high correlation between number of honorific awards and rank of department. Only twenty-one men in the lowest-ranking departments had received more than one award. Statistically rare situations are sometimes theoretically critical.

Table 3. Geographical location of the observed and observers

Region of 120 observed	Mean visibility score and standardized mean awareness score — Region of the 1,308 observers							
	East	Total no. of identifications	South	Total no. of identifications	Midwest	Total no. of identifications	Far west	Total no. of identifications
East	50	(5,174)	44	(2,278)	48	(4,136)	47	(1,637)
South	18	(928)	19	(378)	22	(699)	17	(306)
Midwest	31	(3,703)	31	(1,587)	38	(2,874)	36	(1,222)
Far west	58	(2,294)	55	(945)	62	(1,759)	65	(699)
Total	43	(12,099)	40	(5,188)	46	(9,468)	44	(3,864)

NOTE: Because of the five forms of the questionnaire, not all physicists ranked the same number of men in each category. For example, the physicists receiving form I ranked 9 physicists from eastern universities and physicists receiving form II ranked 15 physicists from eastern universities. For this reason mean awareness scores would not be comparable. An identification consists of one observer saying whether he was familiar with the work of one observed physicist. Thus, each observer made twenty-four identifications. The scores were standardized by taking the total number of possible identifications in each group and dividing this into the actual number of men identified. This standardized awareness score is in fact the same number as the mean visibility score, emphasizing that visibility and awareness are analytically distinct aspects of an empirically unitary process.

it was produced. After all, these men might have greater opportunities to meet one another at regional professional meetings and to read each other's work in regional journals. The data presented in table 4 enable us to reject this hypothesis. Low-quality work is less visible but, again, is no more likely to be visible in the section of the country where it was produced than in other sections of the country.[13] Table 4 also indicates that the low visibility of southern scientists in all sections of the country is not a result of where they work, but of the quality of their work. Southern physicists doing high-quality work are just as visible in all sections of the country as are their nonsouthern colleagues doing high-quality research.

This type of analysis, in which we simultaneously vary characteristics of the observed and observers, enables us to ask two questions. What types of physicists are more visible to some segments of the community than others? What types of physicists have greater awareness of some men than others? To answer these questions thoroughly we would have to present a large number of complex tables. Instead, we present one table that summarizes the findings. Dividing the 120 observed physicists by their visibility scores makes it possible to summarize the findings for all those variables which are highly associated with visibility: quality of work, honorific awards, and rank of department. In the summary table, we characterize the 1,308 observers by the variable we are most interested in, rank of department. We hypothesized that the highly visible physicists would probably be visible to men in all departments, and that the physicists with low visibility would be uniformly unseen. But what of those physicists with medium visibility? Here we might find that location at top-ranked departments increased awareness. The data are presented in table 5. Again we must reject the hypothesis. Physicists are almost uniformly visible to all sectors of the community. Rank of department still fails to make for large differences in awareness, even when the objects being observed were divided by their visibility scores. *No matter how we classified the 120 observed physicists and the 1,308 observers, we found the same results: a man's visibility was equal in all segments of the community, and there were only slight differences in awareness of*

13. Physicists from far-western universities who have produced "low quality" work are more visible than their counterparts from other regions. We suspect that this is due to the fact that some of the physicists in our sample from the far west have relatively few citations to their work but have gained visibility through administration of important laboratories.

Table 4. Quality and region of the observed and region of the observers

Quality of 120 observed	Region of observed	Mean visibility score and standardized mean awareness score — Region of the 1,308 observers							
		East	Total no. of identi-fications	South	Total no. of identi-fications	Midwest	Total no. of identi-fications	Far west	Total no. of identi-fications
High	East	67	(2,928)	59	(1,269)	63	(2,316)	66	(913)
	South	50	(103)	67	(43)	65	(84)	64	(33)
	Midwest	45	(1,785)	42	(776)	50	(1,439)	55	(580)
	Far west	65	(1,479)	64	(597)	72	(1,118)	72	(446)
Low	East	29	(2,246)	26	(1,009)	27	(1,820)	22	(724)
	South	19	(825)	12	(335)	17	(615)	11	(273)
	Midwest	17	(1,918)	20	(811)	27	(1,435)	18	(642)
	Far west	46	(815)	40	(348)	46	(641)	53	(253)

Table 5. Visibility of observed and rank of department of observers

Visibility of 120 observed	Mean visibility score and standardized mean awareness score — Rank of department of 1,308 observers							
	Distin-guished	Total no. of identifications	Strong	Total no. of identifications	Good, ade-quate plus	Total no. of identifications	Fair, poor	Total no. of identifications
High	86	(2,007)	84	(1,906)	80	(3,078)	77	(2,463)
Medium	47	(2,032)	44	(1,944)	44	(3,081)	37	(2,584)
Low	16	(2,567)	15	(2,486)	15	(3,969)	13	(3,185)

scientific research. The data lead to the conclusion that, in physics, location in the stratification system has little influence on the receipt of this type of information.

There was only one variable on which we might expect these conclusions to be reversed, and that is specialty. For it is plausible that even in an open communication system, men working in one specialty will be more familiar with work in that specialty than they are with work in others. Table 6 presents data for the four largest specialties: atomic and molecular, elementary particles, nuclear, and solid state. As we would expect, the work of physicists in each of these specialties is more visible to their colleagues in the same specialty than to colleagues outside it.[14] This is true even when we control for the quality of work. (This may be seen by looking across the rows of table 6.) The differences, however, are considerably smaller than expected. For example, solid-state physicists who have produced high-quality work are only slightly better known to other solid state physicists than to atomic and molecular physicists. This indicates that in physics the possession of this type of knowledge easily permeates the boundaries of specialty. High-quality work done in one specialty is known to almost all physicists regardless of their specialty.

Now let us turn to the significance of the data presented in table 6 for the analysis of awareness. Here the pattern of findings is not clear. Only in three of the eight comparisons do the data go in the expected direction, that is, with physicists having greater awareness of work done in their own specialty. (This may be seen by looking down the columns of table 6.) But even here the differences are not great, and we must conclude that physicists in any one specialty have relatively high awareness of high-quality work in all specialties. We must remember, of course, that we are not in any sense measuring detailed knowledge of the work but rather self-reported familiarity with it.

We have concluded that awareness is not greatly influenced by physicists' individual or contextual characteristics. It is high in all sectors of the community and is taken as evidence that location in the stratification system has little effect on receipt of this type of knowledge.[15] This is not surprising since the formal communication

14. An exception is found in nuclear physics, where the work of nuclear physicists is more visible to physicists working in elementary particles than to those in nuclear physics.

15. A serious difficulty in measuring awareness is that of determining the extent to which the physicists exaggerated their knowledge. As a crude gauge of the validity

Table 6. Specialty of observed and specialty of observers

		Mean visibility score and standardized mean awareness score							
		Specialty of 1308 observers							
Quality of 120 observed	Specialty of observed	Atomic and molecular	Total no. of identifications	Elementary particles	Total no. of identifications	Nuclear	Total no. of identifications	Solid state	Total no. of identifications
High (60 + citations)	Atomic and molecular	63	(274)	54	(807)	50	(626)	51	(700)
	Elementary particles	54	(218)	81	(594)	68	(574)	54	(528)
	Nuclear	61	(265)	70	(759)	57	(570)	48	(597)
	Solid state	42	(138)	35	(390)	36	(289)	49	(289)
Low (0 – 59 citations)	Atomic and molecular	25	(20)	11	(64)	15	(46)	21	(58)
	Elementary particles	36	(246)	59	(684)	41	(542)	26	(554)
	Nuclear	24	(371)	41	(1,040)	38	(837)	23	(911)
	Solid state	12	(133)	11	(408)	15	(291)	23	(261)

system is open to all. The main instrument of formal communication in science is scientific journals. We asked the physicists in our sample to list the three journals that they read most frequently. The

---

of response, a total of five fictitious names was included in the list of physicists, one on each form of the questionnaire. The number of physicists reporting that they were familiar with the work of one of the fictitious men was 51, or 4 percent of the sample. Furthermore, of these 51, all but 3 were reporting on fictitious names which resemble closely the names of actual physicists.

The small differences found on awareness could conceivably be the result of differential tendencies to exaggerate knowledge or systematic differentials in the criteria the physicists used in determining whether they were familiar with the work of a particular man. For instance, if physicists who had produced low-quality work used less stringent criteria in identifying others than their colleagues who had produced high-quality work, this might account for the small differences in awareness between these two groups. The data do not support this interpretation. There were no systematic differences in exaggeration. Four percent of physicists who had published high-quality work, medium-quality work, and low-quality work identified a fictitious name. Four percent of physicists from top-ranked departments, and 4 percent of physicists from poor departments, said they were familiar with the work of a fictitious man. Since exaggeration was limited in general, and since we found no correlates of exaggeration among the variables in our data, it is probable that exaggeration was the result of randomly distributed error.

Our questionnaire also contained a rough gauge of the extent to which there was exaggeration of knowledge of awards. We included one fictitious award on each form of the questionnaire. (One was called the "Richard Saunders Gold Medal—Canadian Roentgenray Society.") Only 6 percent of the physicists, or 79, ranked one of the fictitious awards. Of these, 49 were rankings of the "Pupin Gold Medal," a fictitious award named after an eminent physicist. Only 9 men identified both a fictitious man and a fictitious award. There were no systematic differences in the extent to which fictitious awards were ranked. For example, 6 percent of men in all quality categories ranked a fictitious award.

To check the validity of our results further, we ran some tables excluding those men who had identified fictitious entries. As may be seen in the table below, the results were not significantly affected by the inclusion of men who exaggerated their knowledge. When the exaggerators are excluded, men who have two or more awards ranked an average of three more awards than those who had no awards. When the exaggerators are included, the same results were obtained.

Number of Awards Held	Mean Number of Awards Ranked (Whole Sample)	Mean Number of Awards Ranked (Exaggerators Excluded)
0	10.8	10.5
1	12.1	11.6
2 or more	13.9	13.3

It is true that the exaggerators had higher awareness scores than the nonexaggerators. For example, the 79 exaggerators on awards ranked an average of 17 awards, the nonexaggerators ranked an average of 11. However, since we were able to find no correlates of exaggeration among the variables in our data, we conclude that exaggeration probably had no significant influence on the conclusions we have reached.

most widely read journal is the *Physical Review,* with 77 percent of the sample reporting that they frequently read this journal. The second most widely read journal was *Physical Review Letters;* 59 percent reported that they frequently read it. No other journal was mentioned by more than 25 percent of the sample.

We thought that those physicists who read the most "important" journals should have greater awareness.[16] It turns out that physicists who listed the two most widely read journals had a mean awareness score of 10.8, whereas those who did not list either of the two most widely read journals had a mean score of 8.8. Although "journal reading" explains only a small amount of the variance on awareness,

Table 7. Journal reading, age, and awareness

	Mean awareness score					
Age of observers	Reads 2 top journals	No. of men	Reads 1 of 2 top journals	No. of men	Reads neither of 2 top journals	No. of men
60 or over	13.2	(18)	11.6	(34)	10.1	(37)
59–40	12.4	(239)	11.4	(223)	9.3	(103)
Under 40	9.8	(414)	9.6	(189)	6.9	(51)

it is more powerful than other variables we have discussed. Men who work in poor departments but read the two top journals have greater awareness than men who work at the distinguished departments but do not read either of these journals. Men who have done low-quality work who read the two top journals have greater awareness than men who have done high-quality work but do not read them. In fact, the combined effect of journal reading and age gives us the greatest differentiation on awareness. Those men who have been active in physics for a long time and who still keep up with new developments by reading the key journals have the highest awareness. Young men who do not read the key journals have the lowest awareness (see table 7).

16. We do not mean to imply that journals which are not widely read are not significant journals. However, there is other evidence that the two most widely read journals in physics also are the ones where the most significant research is generally published. See M. M. Kessler, "The M.I.T. Technical Information Project," *Physics Today,* 18 (March):28–36; Eugene Garfield, "Citation Analysis as a Tool in Journal Education," *Science* 178 (November 3, 1972):471–79.

COMMUNICATION OF SCIENTIFIC INFORMATION

The analysis thus far leads to the conclusion that communication of information in physics is relatively open. However, as we have pointed out, the data presented deal only with the communication of the broadest type of knowledge. We have not distinguished between degrees of knowledge that scientists have of each other's work. Although scientists variously located in the stratification system may have been equally likely to be aware of the high-quality work of their colleagues, the actual degree of knowledge may vary substantially. We have no measure of degree of knowledge, but we can easily measure actual utilization. Presumably, to make use of an idea in one's own work requires a greater degree of knowledge than merely to say that one has heard of the idea. In considering utilization patterns we can raise the same questions that we have considered in the analysis of visibility and awareness. Does the location of a scientist in the stratification system influence the type of scientist who utilizes his work? Does the location of a scientist in the stratification system influence the type of work he utilizes?

The data generated to answer these questions were taken from the sample of 120 university physicists. We took a subsample of eighty-four physicists and analyzed the citations made by these men in their paper most heavily cited in the 1965 *SCI*.[17] We then collected information on a one-third random sample of the scientists who were cited in the best paper of each of the eighty-four physicists.[18] These two data sets allow us to see the ways in which location in the stratification system influences the utilization of a scientific work. All the relevant data are presented in a table of zero-order correlation coefficients (see table 8). In the columns of table 8 we have the characteristics of the cited authors. In the rows we have the same characteristics of the citers. One glance at the table tells us that, without exception, the characteristics of cited authors and their citers are related, although none of the correlations are high. Let us now examine the significance of these findings.

The results presented in table 8 are relevant for two problems.

17. For a full description of this sample and the data collected on it, see Appendix A. Our original sample consisted of 120 scientists. Some of these were never cited in the *SCI;* a few had produced only review papers, which were excluded from our sample. Subtracting these nonproducers from the 120, left us with 84 citers for whom sufficient information was available for inclusion in this subsample.

18. For a complete account of the difficulties of collecting data on these physicists, see chapter 8, fn. 6 and 7.

The first problem we consider is how location in the stratification system influences the type of work a scientist utilizes in his own work. Highly ranked scientists are slightly more likely to use high-quality work produced by eminent scientists than are scientists with lower rank. Thus, scientists at top-ranked departments are more likely to use high-quality work than those located at less prestigious institutions. There are several possible explanations of this finding. It is possible that eminent scientists have more detailed knowledge of the work of other eminent scientists. What is more likely, however, is that high-quality work done by other eminent scientists is more relevant for their work. One difference between eminent and noneminent scientists could be the types of problems they choose to work on. Having an eye for the important problem con-

Table 8. Correlation coefficients relating characteristics of authors of scientific papers and characteristics of the scientists who use the papers

	Characteristics of cited authors			
Characteristics of citers	Rank of department	Prestige of highest award	Number of awards	Quality of work
Rank of department	.07	.16	.15	.11
Prestige of highest award	.16	.16	.15	.12
Number of awards	.18	.17	.18	.18
Quality of work	.17	.23	.16	.16

NOTE: There were 385 cited physicists and 84 citers. Each correlation coefficient is based on an N of 385.

tributes to eminence. Thus, high-ranked scientists are probably working on similar problems.

Turning to the second problem, the correlation coefficients of table 8 tell us that eminent physicists are slightly more likely to have their work utilized by other eminent physicists than are their less eminent colleagues. Location in the stratification system explains little variance on who utilizes a scientist's work, but the small correlations are nonetheless quite consistent. These zero-order correlations alone are insufficient to determine how the communication of ideas is influenced by the stratification system. Eminent scientists may utilize the work of other eminent scientists because that work is of higher quality and more relevant to them. We must examine the influence of location in the stratification system on utilization, independently of the quality of work.

In table 9, we present a matrix of partial correlations. In each case, we are looking at the effect of the cited author's location in the stratification system on who cites him, independently of the quality of his work. In general, if we compare the partial correlations of table 9 with the zero-order correlations of table 8 we see that in each case the partial is lower. Thus, part of the effect of location in the stratification system on who utilizes work is a result of the quality of that work. Eminent scientists are more likely to use the work of other eminent scientists because that work is of higher

Table 9. Partial correlations relating characteristics of cited authors and citers, controlling for quality of cited work

Characteristics of citers	Characteristics of cited authors: Rank of department	Prestige of highest award	Number of awards
Rank of department	.04	.12	.11
Prestige of highest award	.13	.12	.11
Number of awards	.14	.10	.11
Quality of work	.14	.18	.10

NOTE: Each partial correlation is based on an N of 385.

quality. The partial correlations in table 9 represent the advantage accruing to a scientist by virtue of his location. Thus, independently of the quality of work, occupying a prestigious position in the stratification system increases visibility and leads to one's work being more widely utilized by other eminent scientists. Let us consider just one concrete example. Of two scientists doing "equal" quality work, the one located at a prestigious academic department is slightly more likely than the one located at a department of lesser prestige to have his work utilized by scientists who have received prestigious awards.

What do these data tell us about the extent to which science is universalistic? Do they offer more support for the conflict or for the functional theory of stratification? Conflict theorists would explain the unequal distribution of citations (utilization patterns) with an "in-group" interpretation. Members of the in-group, who control the journals and major institutions of science, cite each other and ignore the equally valuable work of outsiders. Functionalists would explain the unequal distribution of citations by differences in quality of the work and the value placed on original contributions. The low partial correlations of table 9 would seem to lend support to the functional theory. Eminence (membership in the in-group) has very little independent effect on who utilizes a scientist's work. Noneminent scientists are just about as likely to cite the work of in-group members as are other in-group members. And conversely, in-group members are just about as likely to utilize the work of the outsiders as are outsiders themselves. Conflict theorists could criticize this interpretation on two counts. First, rather than look at the partials of table 9, they would argue that we should look at the zero-orders of table 8. By controlling for citations we are merely partialling out the advantage gained through membership in the in-group, since being heavily cited is a result of in-group status. However, even the zero-orders of table 8 are relatively small, indicating that in-group membership has some advantage but not a great deal. Second, conflict theorists would argue that the reason why outsiders cite insiders is because the in-group controls the means of communication and manipulates the value system; outsiders desiring to be upwardly mobile accept the values of the "ruling" group. The only problem with this interpretation is that it fails to explain why outsiders cite the work of nonelites almost as frequently as nonelites themselves. Although, given these data, there is no way to conclusively settle the issue, the data would seem to offer only scanty support for the conflict interpretation.

THE AWARENESS OF SCIENTIFIC AWARDS

So far we have reported data only on physicists' awareness and utilization of the research of other physicists. We wanted to see if there would be socially structured differences in the communication of other types of knowledge. We were particularly interested in knowledge of the reward system. The questionnaire mailed to university physicists asked them to rank a series of thirty awards and prizes. Two categories of answers were included which indicated that the responding physicists did not have enough information to rank the award or had never heard of it. (For the exact wording of the question, see Appendix A.) A physicist's awareness of awards score was the number of awards he ranked. We can use this score as a measure of knowledge of the reward system. The distribution of awareness scores is presented in table 10. The majority of physicists could identify about one-third of the awards on their questionnaire.

Table 10. Awareness of awards scores

Number of awards known		Number of scientists
0–5		116 (9%)
6–8		225 (17%)
9–13		586 (45%)
14–17		243 (19%)
18–29		124 (10%)
	Total	1,294 (100%)
	NA	14
	Total	1,308

Mean = 11.36
Standard deviation = 4.88
Coefficient of variation = 43

Correlation coefficients between awareness of awards and several variables we thought might influence awareness are presented in table 11. These correlations indicate only small structural differences in awareness of awards. A total of only 10 percent of the variance on knowledge of awards can be explained by the variables included in the study.

Age and professorial rank were moderately correlated with awareness of awards.[19] Both variables have an independent effect

19. It is possible that the relatively low correlations of table 11 may not tell the whole story on awareness of awards. Seemingly low correlations between various attributes of observers and awareness may be masking significant differences in awareness of specific awards. For example, age may be positively correlated with awareness of one set of awards and negatively correlated with awareness of another set, thus creating a small zero-order correlation. Also, "ceiling" and "basement" effects may, in part, contribute to the low observed correlations. For those awards that are either known to almost all physicists or to none, there can be few systematic differences in awareness. There can be no structural bases of awareness of the Nobel Prize, since 100 percent of our sample knew it. Nor can there be extensive differences in knowledge of the National Academy or the Fermi Prize, known to al-

on awareness of the reward system. Among full professors age still counts; those over fifty had an average awareness score 2.2 points higher than men under forty, and, at all age levels, full professors had higher scores than their junior colleagues. Exposure seems to count when observation of awards is considered. Full professors are more apt to know of awards since they are at times asked to review candidates for various prizes and are themselves recipients of prizes, while older scientists, regardless of rank, have been exposed to information about awards for a longer period of time.

Table 11. Coefficients of correlation between awareness of awards and its possible determinants

Independent variable	Correlation coefficient with awareness of awards
Age	.24
Number of honorific awards held	.23
Professorial rank	.20
Prestige of highest award	.17
Quality of scientific output (total number of citations)	.15
Rank of current department	.02
N = 1,294	
NA 14	
Total 1,308	

Award holders

One of our working hypotheses in the study of awareness of awards held that those physicists who were eminent and had received prestigious awards would have a greater knowledge of this aspect of the reward system than those who had not been so honored. This hypothesis is substantially confirmed by the results reported in table 12. Scientists who have received awards have greater awareness of this aspect of the reward system than those who have not. While the number of men who have received numerous awards is small, as expected, these data suggest that high awareness is likely to be an attribute of lofty statuses that are occupied by very few scientists. Elite physicists with seven or more awards had an awareness score almost twice that of scientists without awards.

most 95 percent of the physicists sampled; or if we view the Bruce Gold Medal, known to only 6 percent of our sample. These ceiling and basement effects may be reducing the strength of the correlations.

This select group of highly rewarded physicists knew almost two-thirds of all awards sampled.[20]

Table 12. Number of awards held by physicists and awareness of awards

Number of awards held by observers	Mean awareness score	Total number of men
0	10.8	(944)
1	12.1	(198)
2	13.1	(86)
3	14.3	(34)
4	13.7	(15)
5–6	14.0	(8)
7–8	18.3	(6)
9 or more	23.0	(3)
	Total	(1,294)
	NA	14
	Total	1,308

The results of table 12 might simply indicate that physicists identified their own awards. That awareness of awards is a result of eminence rather than of merely possessing many awards, can be shown by controlling for the prestige of highest award (table 13). The data of table 13 show that it is not the number of awards that determines the scientist's awareness; rather it is the prestige of the awards that he holds that affects his knowledge. The prestige of a scientist's highest award is an excellent indicator of his location in the stratification system. Those who hold highly prestigious awards occupy, for the most part, lofty statuses, and are thus in structurally good positions to know more about formal honors than their less eminent colleagues. These data allow us to conclude that differentials in awareness among scientists who received varying numbers of awards are not a result of those who have won awards identifying their own. This seems apparent since laureates and academy members who had received no other prizes had aware-

20. Since the data indicate the occupants of high-status positions have greater awareness of awards, we examined the simultaneous effect of two of these characteristics: high academic rank and number of awards. The data indicate that the number of awards received is a much stronger determinant of high awareness than professorial rank. At all but one level of number of awards received by physicists, the relationship between professorial rank and awareness has been reduced.

Table 13. Number of awards and prestige of highest award of observers by awareness of awards

	Mean awareness score Prestige of highest award							
No. of awards held by observers	Nobel Prize/ National Academy (4.0–5.0)	Number of scientists	Moderately esteemed (3.0–3.99)	Number of scientists	Less esteemed (0.–2.99)	Number of scientists	Zero awards	Number of scientists
0							11.0	(944)
1	18.0	(2)	12.1	(43)	12.1	(153)		
2	15.0	(3)	13.1	(21)	13.0	(62)		
3	21.5	(2)	12.9	(11)	14.3	(21)		
4 or more	17.0	(18)	15.3	(6)	12.4	(8)		
								(1,294)
							NA	14
							Total	1,308

ness scores just as high as those of scientists with four or more awards.

High awareness of the formal reward system would seem to be an attribute of those who in one way or another are deeply involved in the evaluative process in physics. These are the "gatekeepers" of science—those who control access to rewards and other significant parts of the institution of science. Many are members of the same societies and committees. They are often, therefore, in communication with one another and are familiar with their peers' own experiences with formal recognition (for these men become the relevant reference group for the other elites). Consequently, they have a far greater awareness of both the structure and operation of the reward system of science.

Status-appropriate responses

To complete the story on awareness of awards, we found that specialty in which physicists work, geographical region in which their departments are located, and the quality of their scientific output made for only small, insignificant differences in levels of awareness of the formal reward system. A continual interest of ours has been in finding a contextual determinant of awareness. Consequently, we have recurringly focused on prestige of a physicist's department. In the case of awareness of scientific research, rank of department made for only small differences. Here we see from the correlation coefficients (table 11) that rank of department appears to add nothing to our knowledge of the determinants of awareness of awards ($r = .02$). These findings converge to suggest that there is an open communication system in physics. However, if we examine the effect of structural location on awareness of specific awards, a number of striking results emerge.

The study included a type of data which have some direct bearing on the effect of social location on awareness of awards. Since the questionnaire included five forms, we were able to include the same awards on two forms with slight variations in the way in which the awards were presented. One of these variations involved presenting an award on one form without the name of the granting institution attached to it. In other words, one cue to an award's identity was withheld from one of the two sets of observers. In this way, we could gauge the effect of the presence of this additional cue on the response pattern. The data show clearly that "knowledge" of awards is affected by the presence of this additional cue. Take, for example, only two awards. The Elliot Medal, granted by the National Academy of Science, was known to 2 percent of the physicists when it appeared without being identified as being granted by the National Academy, and to 14 percent when it was so identified. Thirty-two percent of the physics community had heard of the Rumford Medal of the American Academy of Arts and Sciences when the award was presented without the institutional cue, and 45 percent with it.[21] More significant than these sheer differentials in visibility was the way in which these differences were distributed among the physicists. Table 14 presents awareness scores of both the Elliot and Rumford medals and the rank of department of observing physicists. In both cases, the added information makes for little difference among members of low-prestige departments, but for a considerable difference among members of distinguished departments.[22]

Table 14. Rank of department of observers and awareness of two awards presented with and without name of granting organization

Award as it appeared on questionnaire	% identifying award — Rank of department of observer: Distinguished departments	Number of observers	Low-ranked departments	Number of observers
Elliot Medal (National Academy of Sciences)	20	(64)	3	(62)
Elliot Medal	–	(63)	3	(71)
Rumford Medal (American Academy of Arts and Sciences)	64	(52)	42	(64)
Rumford Medal	33	(63)	34	(71)

21. The Rumford Medal appeared on two forms of the questionnaire. On the first form the additional cue was present. The Elliot Medal appeared on two forms—one with the added cue and one without it.

22. In tables 14 and 15 we compare awareness of awards only for those scientists located in the highest-ranked departments ("distinguished") and in the lowest-ranked departments. We compare extremes because these are the theoretically interesting cases. The fact that part of the sample is left out of table 14 accounts for the difference between the percentage identifying the Rumford Medal reported in the text and the figures reported in the table.

One possible interpretation of these data utilizes the notion of a "halo-effect." This explanation holds that evaluations are influenced by the institutional context of the object being judged. Accordingly, scientists who work at Harvard have higher visibility than men who produce work of equal "quality" at lesser universities. Similarly, of two academic departments of equal quality, the one located in the more prestigious university will be rated higher. Following this line of reasoning, scientists who associate prestige with organizations such as the National Academy are more likely to know of awards, or think they know of awards, granted by the Academy. But if this explanation is correct, we still must explain why the halo-effect is unevenly distributed, why it has a greater influence on scientists located at prestigious departments than their colleagues in low-prestige departments.

A plausible interpretation of the data of table 14 employs the concept of "status-appropriate response." Individuals often express themselves in a particular way because the tradition or norms of their social status require a particular type of behavior. In fact, norms related to individual statuses may lead the occupants to give a particular impression even when the status occupant is not consciously disposed to create such an impression.[23] Status-appropriate responses apply not only to expected modes of behavior but also to expected knowledge associated with social positions. It is possible that scientists located at high-prestige departments may feel obliged to be familiar with information of all kinds that someone further removed from the center of scientific activity would not know or be expected to know. For men associated with the most prestigious institutions, there is a belief that it is appropriate to know of people, research, and even awards associated with other highly prestigious organizations.

Scientists at distinguished departments may feel that they should know of those awards given by prestigious societies such as the National Academy of Sciences or the American Academy of Arts and Sciences. Status-appropriate responses could account for the specified results obtained when we looked at the effect of additional cues on the identification of awards. It follows, as well, that there are status-appropriate responses among men of different profes-

23. An interesting and informative discussion of this point can be found in Erving Goffman, *The Presentation of Self in Everyday Life* (New York: Doubleday Anchor Books, 1959), Introduction.

sorial ranks, different ages, and different levels of scientific eminence.

If the interpretation is sound, we should find the sharpest differentials in the distribution of knowledge for awards granted by prestigious organizations. Awards given by bodies of less prestige should not elicit a similar differentiated response between men at departments of differing rank. Table 15 presents data in support of this hypothesis. For awards granted by highly prestigious organizations, such as the National Academy, Atomic Energy Commission, and Franklin Institute, awareness is increased far more by the additional cue among members of distinguished departments than among members of low-prestige departments. Corroborative evidence lies in the fact that there are virtually no specified differences in effect for awards granted by less prestigious bodies such as the American Association of Testing and Materials or the American Association of Physics Teachers. In other words, while physicists at distinguished departments feel it appropriate for men in their position to know of awards given by highly visible and esteemed organizations, they do not feel it necessarily appropriate to know other awards granted by less notable societies. The only exception is the Enrico Fermi Prize of the Atomic Energy Commission. This anomaly might be explained by the fact that this award has such high prestige and visibility that little difference could result from additional cues.

CONCLUSION

It is clear that characteristics which make for high visibility of research do not make for high awareness. Quality of work, possession of honorific awards, rank of department, and specialty combine to explain 61 percent of the variance on visibility. The same variables, however, explain only a small amount of the variance on awareness of research. This is the case for physics. It remains to be seen whether these conclusions hold for other scientific disciplines and even the humanities. This research has been a case study of a single science. It seems probable that different results might obtain in disciplines that are not as highly institutionalized as physics. Compared to some other fields, physics has been fractionalized less often and less severely. In physics, there may be greater consensus on what is or is not significant than in many other fields. We would expect that, in less institutionalized fields, location in the stratifica-

Table 15. Rank of department of observers and awareness of awards presented with and without name of granting organization

Awards as appearing on two forms	% identifying award					
	Distinguished departments	N	Difference	Low-ranked departments	N	Difference
Rumford Medal (Premium) (American Academy of Arts and Sciences)	64	52	31	42	64	8
Rumford Medal (Premium)	33	63		34	71	
Elliot Medal (National Academy of Science)	20	64	20	3	62	0
Elliot Medal	0	63		3	71	
Frederic Ives Medal (Optical Society)	30	216	20	24	274	8
Frederic Ives Medal	10	63		17	71	
Comstock Prize (National Academy of Science)	19	64	13	8	62	5
Comstock Prize	6	63		3	71	
Elliot Cresson Medal (Franklin Institute)	15	216	12	8	274	4
Elliot Cresson Medal	3	63		4	71	
Ernest Orlando Lawrence Award (Atomic Energy Commission)	75	57	7	52	71	0
Ernest Orlando Lawrence Award	68	63		52	71	
Eugene Cook Bingham Medal (Society of Rheology)	12	52	7	14	64	11

Eugene Cook Bingham Medal	5	63		3	71	
Bruce Gold Medal (Astronomical Soc. of Pac.)	8	64	6	—	62	−3
Bruce Gold Medal	2	63		3	71	
Capt. Robert Dexter Conrad Award (Office of Naval Research)	6	52	1	2	64	1
Capt. Robert Dexter Conrad Award	5	63		1	71	
Dannie Heinemann Prize (American Association of Physics Teachers)	67	52	−1	53	64	1
Dannie Heinemann Prize	68	63		52	71	
Enrico Fermi Prize (Atomic Energy Commission)	94	215	−1	88	268	3
Enrico Fermi Prize	95	63		85	71	
Dudley Medal (American Association of Testing and Materials	2	64	−1	5	62	1
Dudley Medal	3	63		4	71	
Oliver E. Buckley Solid-State Physics Prize (American Association of Physics Teachers)	69	52	2	72	64	13
Oliver E. Buckley Solid-State Physics Prize	67	63		59	71	

NOTE: This table does not present data for members of strong or fair-to-good departments. Here we examine extreme categories.

tion system might make for greater differences in awareness. We must also point out that we have been dealing solely with physicists working in universities offering Ph.D.'s in physics. This, of course, is a relatively elite population. It is likely that had we sent out questionnaires to physicists teaching at undergraduate colleges and to those working in industry, we would have found greater differences in awareness.

When we examined the actual utilization of research, we found results similar to those on awareness. Location in the stratification system had little influence on either what work a scientist utilizes or who utilizes his work. Perhaps most important about this part of the analysis was that the data offer little support for an "in-group" interpretation of citation practices. Elites cite the work of nonelites as frequently as nonelites themselves.

Finally, we examined in detail another aspect of how location in the stratification system influences receipt of knowledge: awareness of honorific awards. There were few patterned differences in awareness of awards. Those bases of awareness of awards that did obtain were attached to lofty statuses. High awareness was found almost exclusively among the most eminent men of contemporary physics, those men who had been honored themselves with prestigious awards. Extensive knowledge of this aspect of the reward system rests in the hands of the "gatekeepers" of science. This points to the necessity of making future qualitative inquiries into structural bases of awareness, looking particularly at those few men who are most intricately involved in the evaluative system.

In this chapter, we have focused on awareness and utilization of knowledge at one point in time. Clearly, these are dynamic rather than static processes. Significant problems arise when we introduce the added variable of time. Particularly, how does location in the stratification system affect the initial reception of ideas, the speed and extent of their diffusion throughout the social system, their resistance or assimilation over time? These problems can only be dealt with when we consider the temporal element in utilization. It is this problem that we consider in the following chapter.

7 PROFESSIONAL STANDING AND THE RECEPTION OF SCIENTIFIC DISCOVERIES

Progress in science depends upon the rate of discovery and the efficiency with which discoveries are evaluated, diffused, and incorporated into the body of scientific knowledge. For this reason, the sociology of science analyzes the social conditions which affect the processes of discovery, evaluation, and diffusion. This chapter presents data from a series of studies of the diffusion of scientific ideas.

We start with the null hypothesis that sociological variables have no influence on the processes of science. Indeed, it is this hypothesis which is probably held by many working scientists. The development of science is seen as the result of the internal dynamics of scientific ideas. If science were comprised of nothing more than the dynamics of ideas, it would be the most rational of human institutions; the evaluation and diffusion of discoveries would depend solely upon their intellectual substance. Neither the personal attributes of the discoverer and his location in the social structure of science nor the similar characteristics of his audience would influence the reception of the discovery.

An alternative hypothesis for these studies was presented by Robert K. Merton. In his paper, "The Matthew Effect in Science," Merton suggested that in science as in other areas of human life those who are rich are likely to get richer.[1] Merton takes the title of his paper from the Gospel according to St. Matthew: "For unto every one that hath shall be given, and he shall have abundance; but from him that hath not shall be taken away even that which he hath." The Matthew Effect consists "in the accruing of greater increments of recognition for particular scientific contributions to scientists of considerable repute and the withholding of such recognition from scientists who have not yet made their mark."[2]

1. Robert K. Merton, "The Matthew Effect in Science," *Science* 159 (January 1968):56–63.

2. Ibid., p. 58.

Merton hypothesized that if two scientists independently make the same discovery, the considerably more eminent one will get the greater or perhaps all the credit. Likewise, if scientists of greatly differing status collaborate, the one who is most eminent will get the lion's share of the credit for the joint effort. Merton performs a functional analysis of the consequences of the Matthew Effect both for individuals and the social system. Since the Matthew Effect involves misallocation of credit, it is dysfunctional for the careers of some individuals; however, it is hypothesized that this same misallocation is distinctly functional for the communication system of science. Since the evaluation and utilization of papers in part depends upon the reputation of the author, discoveries made by eminent men or having eminent men as coauthors are more likely to be quickly incorporated into the body of scientific knowledge. "It leads us to propose the hypothesis that a scientific contribution will have greater visibility in the community of scientists when it is introduced by a scientist of high rank than when it is introduced by one who has not yet made his mark."[3]

It is the purpose of this chapter to test empirically the "Matthew Effect" hypothesis. If science is truly universalistic, then the rank of the author in the stratification system should explain little variance on the initial reception of new ideas. Although Merton implies that the Matthew Effect applies to all scientific work, he limits his analysis to cases of multiple discoveries and collaboration. He did this because a test of the Matthew Effect is valid only if the reception of work of equal quality is considered. In a multiple discovery the work of all participating authors is of roughly equal caliber; in collaboration, two or more authors are producing the same discovery. When scientific papers are of unequal quality, it cannot be said to what extent differences in reception of these papers result from the position of their authors in the stratification system or from the intrinsic qualities of the papers. For this reason, an ideal test of the Matthew Effect would use multiple discoveries. However, multiple discoveries are difficult to identify, and the concept of the Matthew Effect would have greater utility should it turn out to apply to all kinds of scientific discoveries. We are interested in the influence of location in the stratification system on all discoveries. Is work of a given quality more quickly recognized and more widely diffused when the author occupies a position of eminence? To test

3. Ibid., p. 59.

the influence of stratification variables we must be able to control for the quality of papers.

By looking at the number of citations to papers, we can roughly gauge their quality. Citation counts must also be used as a measure of the speed and extent of a paper's diffusion. In all the studies reported here we have begun by counting the number of citations received by a sample of papers in the 1966 *SCI*. By controlling for the number of citations received in 1966, we can get groups of papers that, several years after publication, have equal impact or are, in our terms, of roughly equal quality. We then look at the citations received by the same papers at an earlier point in time, a point closer to their publication. These earlier citations tell us the extent to which papers that were utilized to the same extent at time 2 were utilized at time 1. The hypothesis is that, if assessed quality at time 2 is controlled, the assessed quality of a paper at time 1 partly depends upon the position of its author(s) in the stratification system. To test this hypothesis, we make use of data drawn from several different studies, all of which are similar in design. All deal with citations to papers at two points in time. One study makes use of a sample of papers published in the *Physical Review;* two others of the research output of a sample of university physicists. Another study makes use of a random sample of papers from several fields of science. All studies will be described in detail as we proceed.

SPEED OF DIFFUSION

The literature of modern science, as Price has pointed out, grows at an exponential rate.[4] In many scientific journals more than 50 percent of the references are to work published within the preceding five years. In a science inundated by new literature, it is important that new ideas be recognized and utilized quickly. If a discovery is not recognized soon after it is made, there is a high probability that it will not be recognized at all. Given the importance of early diffusion, we begin the analysis with data from a study of the immediate reception of papers published in the *Physical Review* in 1963. We took a two-thirds sample of these papers and looked them up in the 1966 edition of the *SCI*. The distribution of citations in 1966 to the 1,187 papers studied is presented in table 1. These statistics, generated from an analysis of the papers published in one of the most prestigious scientific journals in the world, suggest that

4. Price, *Little Science, Big Science.*

most papers have little or no visible impact on papers published at a later time.[5] In the third year after their publication, 29 percent of the papers were not cited even once, and 85 percent received five or fewer citations.[6] We decided to study the reception of those papers which have a relatively high impact, the 15 percent cited six or more times in 1966. These 177 papers received an average of ten citations in 1966 and an average of six citations in 1964, the year after their publication.[7] The correlation between the number of citations received by the paper in 1966 with the number received in 1964 was high (r = .72). This indicates that papers receiving heavy use three years after publication were also likely to receive relatively heavy use in the first year after publication. The correlation, however, is not perfect; some papers that were deemed useful in 1966 were less heavily used or went unnoticed in 1964.[8] If the

Table 1. Distribution of citations in 1966 to papers published in the *Physical Review* in 1963:

N Citations in 1966	N Papers	Cumulative %
0	342	29
1	216	47
2	173	62
3	128	72
4	89	80
5	62	85
6 or more	177	100

5. The *Physical Review* is not only the most prestigious physics journal, it is also the most widely read in the field (see chap. 6).

6. It should be clear that we are dealing with citations in a single year: 1966. Some of the papers not cited in 1966 may have been cited in 1965 or 1964.

7. Throughout the research reported in this chapter we have faced a number of difficult methodological problems. One of these has been the increase in the number of journals covered by the *SCI*. Since its inception in 1961, the *SCI* continued to add journals from which it indexed citations. Thus, it is possible that an increase in citations to a paper between 1961 and 1966 might not be due to this paper being more widely used but, rather, to an increase in the *SCI*'s file. In order to assess the effect that increase in file would have on the results, we controlled for this in one study; we counted only citations in 1966 that appeared in journals included in the 1961 file. Since we found that the citations due to increase in file were randomly distributed, with the work of all groups of scientists benefiting equally from the increase, we concluded that it was unnecessary to control for this increase in the other studies.

8. There were hardly any cases in which a paper received more citations in 1964 than in 1966.

Matthew Effect were in operation, we would expect that, when the variance due to the number of citations a paper received in 1966 is removed, the location of the authors in the stratification system of science would affect the initial reception of the papers in 1964. If the response to papers were solely on the basis of their scientific content, there should be no systematic differences in the reception of the work of eminent scientists and noneminent scientists, young and old, scientists in the top departments and those in the less prestigious departments.[9] The data are presented in table 2.[10]

In line A of table 2 the ranking authors of the 177 papers are classified by the number of citations their other papers received in 1964. The number of citations to other papers may be taken as an indicator of repute based upon scientific accomplishment. Those physicists whose other work had received a large number of citations had earned a widespread reputation as a result of their past scientific success. Those whose other work had received few citations had achieved a very limited or no scientific reputation based upon their scientific accomplishment. These data indicate

9. Perhaps a more serious methodological problem than the increase in *SCI*'s file was the problem of multiple authorship of the papers studied. In the study reported here, data were collected on all the authors of each paper. However, since the authors of a given paper often came from different levels of the stratification system, any variance due to stratification might be washed out by using the author as the unit of analysis. Therefore, the paper was used as the unit of analysis; each paper was classified as if it had been written by the ranking author, that is, the one having the highest position. This was determined by the number of honorific awards listed after a man's name in *AMS* and whether he was a Fellow of the APS. In cases where there were no differences in the rank of collaborators, the first author was treated as the ranking author. In the other studies reported in this chapter, we have been unable to collect data on all authors of papers and have therefore treated each paper as if it were written by the first author. The results of these studies are confirmed by the similar findings in the study which treats all the authors of a paper. A more complete testing of the hypotheses would, of course, require the collection of data on all the authors of papers included in the sample. The information on the author's characteristics were collected from *AMS*. In the case of academic scientists, the rank of their department was based upon ratings reported in Cartter, *An Assessment of Quality in Graduate Education*.

10. Before beginning the analysis of the data in this table, the difference between the way the Matthew Effect is being used here and the way it is used by Merton should be clarified. Merton used the Matthew Effect to describe the consequences of the sharpest differences in rank. Most of his examples are of Nobel laureates, the most eminent scientists, collaborating with or being partner to a multiple discovery with a student or other scientist of low visibility. In generalizing the concept of the Matthew Effect, we are using it to describe the effect on the diffusion process of stratification on all levels. Although all the data in table 2 are relevant for a test of the Matthew Effect in its more general sense, only some of them, specifically those dealing with sharp differences in rank, are relevant for an analysis of the Matthew Effect as the concept was used by Merton.

that scientific reputation based upon past performance does have some influence on the reception of new discoveries. The more citations a physicist's other work had received in 1964, the greater was the probability that a new, significant discovery would be immediately recognized (that is, would receive a large number of citations in 1964, a year after publication). The zero-order relationship between the number of citations in 1964 to the other work of the physicist and the number of citations in 1964 to the paper under consideration is $r = .23$.[11] In order to eliminate variation due to differences in the quality of these papers (the number of citations they received in 1966), we computed the partial r. Since the partial r in line A of table 2 is .18, we may conclude that a man's reputation based upon past published work did have a slight independent effect on the reception of a new paper.[12]

Table 2. Correlation and partial correlation coefficients between early recognition of physics papers and several stratification variables (*Physical Review* sample)

Stratification variable	Correlation with early recognition	Partial correlation with early recognition, controlling for quality
A. Scientific repute (N citations in 1964 to all other papers)	.23	.18
B. Rank of academic department (1963) *	.21	.18
C. Membership status in APS (1963)	−.04	.00
D. N honorific awards (1963)	.13	−.04
E. Prestige of highest award (1963)	.12	.01
F. Age (1963)	.04	−.02

N = 177

NOTE: Early recognition is measured by the number of citations the paper received in 1964. Quality is measured by the number of citations the paper received in 1966.

* All authors not working in academic departments were excluded from the analysis (N = 85).

11. For a hypothesis as to why this correlation coefficient was not higher, see note 14 below.

12. The small partial correlations of table 2 and the other partial correlations throughout the chapter could be a result of differential regression effects. If we look at the papers in any group which receive a large number of citations at one point in time, they are likely to regress towards the mean at a second point in time. If rank in the stratification system and citations to papers at two points in time are all con-

Another variable which had a slight independent effect on speed of diffusion was the scientist's institutional location: line B of table 2 indicates that men at prestigious academic departments are the most likely to have their work immediately cited. Tabular treatment of the data showed that institutional location made little difference among those with more than fifty citations to their other work; among those with fewer than fifty citations to other work, the physicists at the distinguished departments still were considerably more likely to have their work quickly recognized. We may conclude that if a physicist had produced work in the past that was currently being heavily utilized, the probability would be very high that a new important discovery by him would be immediately recognized regardless of where he worked. If a physicist had not produced important work in the past, his chances of having a new discovery immediately recognized would be slightly better if he worked at a high-prestige academic department. These latter physicists are aided by their location at strategic points in the social system of science; they are more likely to be tied into the informal communication system of their discipline.

For line C of table 2 we divided the ranking authors of the 177 papers into those who are Fellows of the American Physical Society (APS) and those who are not. Fellows are elected for having made an original contribution to physics; only 10 percent of the 25,000 APS members hold this honor.[13] The data indicate that Fellows are no more likely to have their papers immediately recognized than are non-Fellows. If being a Fellow is taken as an indicator of some degree of eminence, we can say that the Matthew Effect does not operate in this situation. This conclusion is supported by the data in line D of table 2. Here we have classified the authors by

---

sidered to be imperfect measures of eminence, then the fact that, of two groups of papers equal at time 2, the one produced by the low-ranking scientists will have fewer citations at T_1, could be a result of a regression effect. Whether the partial correlations are wholly or in some part a result of regression effects, depends upon how much measurement error there is in the variables. Since all the studies we report on here lead to the conclusion that the Matthew Effect explains very little variance on the reception of scientific papers, the conclusion would only be strengthened if the small partial correlations resulted from differential regression. For a discussion of regression effects see Donald T. Campbell and Keith N. Clayton, "Avoiding Regression Effects in Panel Studies of Communication Impact," *Studies in Public Communication* 3 (Summer 1961):99–118. We thank Lowell Hargens for pointing this problem out to us.

13. Whether a physicist was a Fellow of the Society was determined by stars appearing before the man's name in the society's membership directory, *Bulletin of the American Physical Society,* ser. 2, vol. 10 (1965).

the number of honorific awards listed after their names in American Men of Science (AMS). Although the zero-order relationship between number of honorific awards and initial reception of the paper is .13, this small correlation is totally due to variation in the quality of the papers. When the variance due to the number of citations the papers received in 1966 is removed, the partial r is –.04. Similar results were obtained when we used the prestige of the ranking author's highest honorific award as the independent variable (see line E of table 2).[14]

The findings on age are also contrary to the original hypothesis. We at first thought that older scientists who were still publishing might be more visible and therefore have their work more immediately recognized than their younger colleagues. It turns out that papers written by men under forty are no less likely to be immediately recognized than those of scientists over forty (line F of table 2). We may conclude that longevity in the field does not enhance one's chances of having papers immediately recognized.

The data of table 2 give us a better idea of the extent to which several aspects of the stratification system of science influence the reception of papers. The most important fact is that the quality of the paper being studied, as measured by the number of citations it received in 1966, exerts the primary influence on whether the paper receives immediate recognition. Thus, science does closely approach its ideal of universalistic and rational evaluation of work. Although scientists are clearly universalistic in evaluating new papers, several aspects of location in the stratification system have a slight influence on a paper's reception. Most important of these is

14. In interpreting these data, we should take into consideration that the sample was not large enough to give us great differences in eminence as conferred through formal awards. Tabular treatment of the data showed that those physicists who were not in *AMS* fared less well than those who did appear. These data seem to indicate that only substantial differences in eminence will affect the speed of diffusion. Those scientists who are not known well enough to be listed in the *AMS* and therefore have very low visibility may have to wait longer for their work to be diffused. At first it may appear odd that we found no correlation between the mean number of citations papers received in 1966 and the eminence (as measured by status in the physical society and number of honorific awards) of their authors. This finding, which goes contrary to those in previous chapters, is probably an artifact of the way in which the sample was chosen. Since only high-quality papers were chosen for the sample, it was unlikely that we would find significant differences in the mean number of citations in 1966 to the papers, no matter how we divided the authors. We are confident that had we included all papers in the *Physical Review* we would have found a correlation between the number of citations the papers received in 1966 and the eminence of their authors.

the extent to which a scientist's past work is being utilized at the time of a new discovery. The Matthew Effect also operates for those people located at prestigious points of the social system of science. Contrary to our hypothesis, however, eminence, at least as measured by the data of lines C and E of table 2, has no influence on the speed of diffusion of a scientist's work.

At this point, we can return to an analysis of the functional consequences of the Matthew Effect for individuals and the communication system of science. This study includes only those discoveries which were at time 2 recognized as significant; there is, of course, no way of identifying significant discoveries which are wholly overlooked as a result of the low visibility of their authors. If any such papers exist, there is a possibility that the authors might be "victims" of the Matthew Effect. Since there are no data on overlooked papers, this must remain conjecture.

What about the papers that were not overlooked? The Matthew Effect clearly involves slight inequities in recognition for work of roughly equal significance. Men, for example, who are not at distinguished departments and whose other work is not widely used, must occasionally wait longer for their discoveries to be recognized. Is the Matthew Effect functional for scientific communication? In the cases of multiple discovery and collaboration, the Matthew Effect is functional for science: nothing is lost, and speed of diffusion is achieved. However, when the Matthew Effect is applied to all types of scientific work, it may be seen as dysfunctional for scientific advance. Consider two papers of equal quality but different subject matter written by authors of different rank. If one is immediately recognized and incorporated into the body of scientific knowledge and the other is ignored, progress will be less rapid than if both significant discoveries are immediately recognized. If we start from the assumption of a model of complete rationality in which all significant discoveries are immediately recognized, then delay in recognition of the work of men of low rank would be dysfunctional. However, if, as we know to be the case, diffusion is partly influenced by the author's characteristics, then we can see the early recognition of the work of high-ranking men as functional. Perhaps most important is the fact that a large majority of significant discoveries are immediately recognized. It is not necessary to explain the immediate recognition of the paper of a high-ranked man. It is more sociologically problematic to explain a delay in the recognition of a paper of equal quality by a low-ranked man.

THE MATTHEW EFFECT IN COLLABORATION

So far the significance of the Matthew Effect in determining the speed of diffusion of scientific discoveries has been examined. Each paper has been treated as if it had only one author.[15] This, of course, is far from the case; the 177 papers had a total of 362 authors. We now address ourselves to the next question: how do the attributes of the coauthors affect the reception of scientific discoveries? The data tell us nothing about the dysfunctions of the Matthew Effect for some of the collaborators, but they do provide an approximate test of the hypothesized communication function of the Matthew Effect in collaboration. If the Matthew Effect is indeed at work, we should find that when a man occupying a low rank in the stratification system publishes a paper with a colleague of substantially higher rank, his work will be more quickly diffused than if his collaborator was of equally low rank. To test this hypothesis, we divided the 362 authors into two groups: those who had high scientific repute on the basis of past work and those who had little or no scientific repute on the basis of past work.[16] (Authors having no collaborators were excluded from the analysis.) Then, within each of these groups, we computed a correlation coefficient between the scientific repute (number of citations in 1964 to other work) of the most highly cited collaborator and the initial reception of the 1963 *Physical Review* paper.[17] We then had to compute partial r's, removing the variance due to the quality of the paper (number of citations in 1966). The results, presented in table 3, offer some support for the alternative hypothesis of this paper.

15. See note 9 above.

16. If this were not done, it would be impossible to find any variation due to repute of collaborators. If a paper had a high- and a low-repute author, they would cancel each other out; the high-repute author receiving no increment in visibility from his low-repute collaborator. The validity of this procedure is supported by the data. When we lumped together the two groups of authors of table 3, we got a zero-order correlation of .09 and a partial r of .05.

17. Here again we face the problem of papers having more than one author. We could only include those papers on which the collaborators were either the only author or the first author, as this is how the *SCI* is arranged. However, we know that the correlation between the number of citations to the work on which a man was the only or first author and the number of citations to all his work was very high—$r = .96$ (see chap. 2). There was an additional problem in classifying coauthors. In cases where an author had more than one collaborator, we decided to classify the author by his most highly cited coauthor rather than adding the citations to the other work of all collaborators. However, there are some cases in which increases in visibility of collaborators might be additive. This would occur when each of the collaborators had different audiences rather than overlapping audiences. A more refined analysis would be necessary to pinpoint the effects on diffusion of various patterns of collaboration.

Authors of little or no scientific repute benefit from having high-repute collaborators. Furthermore, this finding is not an artifact of differences in quality of the papers. Although the correlation coefficient is not very high, the Matthew Effect does aid the diffusion of the work of scientists who have not yet acquired a reputation on the basis of past work. When we classified the collaborators by other variables such as the number of honorific awards they have received, we found no evidence of the presence of the Matthew Effect. This was to be expected; if honorific awards would not aid the diffusion of one's own work, they were unlikely to aid the diffusion of the work of one's collaborators.

Table 3. Correlation and partial correlation coefficients between early recognition of physics papers and the scientific repute of the author's collaborators (*Physical Review* sample)

Scientific repute of author	Correlation between N citations in 1964 to other work of most highly cited coauthor and early recognition	Partial correlation between N citations in 1964 to other work of most highly cited coauthor and early recognition, controlling for quality
Low (0–19 citations to other work)	.24	.24
High (20 or more citations to other work)	−.02	−.02

NOTE: Early recognition is measured by the number of citations the paper received in 1964. Quality is measured by the number of citations the paper received in 1966.

EXTENT OF DIFFUSION

So far we have limited the analysis to short-term diffusion of papers published in one journal. These data may suggest that science approaches its ideal of rationality more closely than in fact it does. In designing a study that deals with a short period of time and with papers in only one journal, we may be "controlling out" the variables which produce divergences from the ideal. It is possible that some articles published in 1963 that will ultimately be deemed useful were not yet recognized in 1966. We might also find that the stratification system has a greater impact on diffusion of ideas when we consider the whole of a man's work published over a longer period of time in many diverse journals. In this section we analyze how the stratification system influences the extent of diffusion of discoveries over a longer period of time.

To begin with, the citation patterns of the work of a random sample of 1,308 university physicists were examined.[18] We were specifically interested in the most highly cited paper published by each physicist between 1950 and 1961.[19] Since we want to study the diffusion of discoveries with relatively high impact, any scientist not having a pre-1961 paper with at least ten citations in the 1966 *SCI* was excluded from the analysis. We now wanted to see the extent to which papers judged to be of roughly equal value in 1966 were utilized in 1961, a point in time closer to their publication. If the Matthew Effect is in operation, we should find that, when quality of the paper is controlled, the higher the rank of an author in the stratification system the more widely diffused his discoveries would be in 1961.

In analyzing table 4 we must again begin by noting that the "quality" of papers (the number of citations in 1966) is a more important determinant of extent of diffusion than any of the stratification variables. The zero-order correlation between the number of citations received in 1966 and 1961 was .69. The zero-order correlations between the stratification variables and extent of diffusion in 1961 ranged from .33 for the repute of the scientist's other work to .17 for the rank of the author's academic department. However, when the variance due to the number of citations to the paper in 1966 (the control for quality) is removed, the partial correlations obtained are sharply reduced. The only two variables that have even a slight independent effect on reception of the papers are the prestige and number of honorific awards. These data on papers from a wide range of journals and over a longer period suggest that the Matthew Effect, in the sense it is used here, had relatively no independent influence on the diffusion of scientific ideas. Those papers which were heavily used in 1966 were also heavily used in

18. For a description of this sample, see Appendix A. This is a subsample of the 1,308 university physicists. We have excluded approximately 100 cases on which we could not obtain *AMS* information.

19. We limited our study to papers published between 1950 and 1961 because we wanted some differences in the age of papers, but we felt it would be too difficult to interpret changes in rates of citations to papers published prior to 1950, as the earliest citation index is for 1961. It turned out that differences in the age of papers did not influence the results.

1961, regardless of the location of the authors in the stratification system.

Although the Matthew Effect did not influence the extent of diffusion of the "best" papers of the university physicists, it might have influenced the extent of diffusion of high-quality work done in other fields of science. We studied all papers (in the various fields of science covered by the *SCI*) which were published between

Table 4. Correlation and partial correlation coefficients between extent of diffusion of physics papers and several stratification variables (university physicist sample)

Stratification variable	Correlation with extent of diffusion	Partial correlation with extent of diffusion, controlling for quality
A. Scientific repute (N citations in 1961 to all other papers)	.33	.02
B. Prestige of highest award (1961)	.24	.10
C. N honorific awards (1961)	.22	.08
D. Age (1961)	.18	.02
E. Rank of academic department (1961)	.17	.03

N = 91

NOTE: Extent of diffusion is measured by the number of citations the paper received in 1961. Quality is measured by the number of citations the paper received in 1966.

1950 and 1961 and received at least thirty citations in 1966.[20] These papers, still heavily cited five to sixteen years after publication, represent a sample of the most significant discoveries of the time in a wide range of scientific disciplines. Would the Matthew Effect make any differences in the diffusion patterns of these ideas?

The stratification variable used was the repute of the scientist's other work as measured by the number of citations this work received in 1961. The data indicate that whether one's past work was

20. In our study of papers receiving thirty or more citations in 1966, we excluded all those authors who could not be located in *AMS*. We did a spot check on a sample of names that we could not locate and found that most of them were either foreign scientists or young scientists whose names had not yet found their way into *AMS*.

being widely used did not affect the extent of diffusion of the paper. Although the zero-order correlation between citations to other work in 1961 and citations to the "super" paper in 1961 was .20, this correlation was reduced to .02 when the number of citations the paper received in 1966 was held constant.[21] The papers of those scientists whose other work received few citations were just as widely diffused in 1961 as the papers of historically equal value published by more heavily utilized authors. For these top papers, the Matthew Effect would seem to have little influence on diffusion. Quality of work is by far the most significant determinant of early diffusion. The zero-order correlation between the number of citations these "super" papers received in 1966 and 1961 was .80.

Since it is often difficult to specify the one paper in which a discovery is presented, and since discoveries are often communicated in a series of papers, we wanted to look at the effect of stratification on the reception of all of a scientist's work. Again, we used the sample of 1,308 university physicists. As in the case throughout the analysis, we were predominantly concerned with the reception of work of relatively high impact. We have excluded from analysis those physicists who received fewer than twenty citations in 1966 to all their work published between 1950 and 1961. The question is, how widely diffused was the same work in 1961, a point in time closer to its publication? The data are presented in table 5 and indicate that the more eminent a physicist was in 1961, the more widely his work was diffused. Prestige of highest award, number of awards, and rank of academic department, all of which were measured as of 1961, are all correlated with the total number of citations received in 1961. Furthermore, these correlations hold up when we take into account the variation due to differences in quality of work (the number of citations received in 1966 by pre-1961 work). These findings contrast sharply with those reported above. While the Matthew Effect had little independent influence on the reception of single papers, it does have some influence on the utilization of the whole body of a physicist's work. Since we know that the Matthew Effect had negligible influence on the reception of the "best" paper of the university physicists (see table 4), we can conclude that the partial r's of table 5 are a result of slight but cumulative citations to the physicist's "lesser" papers. We reach the tentative conclusion that the reception of top papers will not be influenced by a scientist's

21. For an explanation of why the zero-order correlation was so low see note 14 above.

position in the stratification system, but that high-ranking scientists are more likely to accumulate citations to their work of relatively small significance. Good papers do not need the Matthew Effect to attain visibility, but less significant papers benefit from it.

Table 5. Correlation and partial correlation coefficients between extent of diffusion of a physicist's work and several stratification variables (university physicist sample)

Stratification variable	Correlation with extent of diffusion	Partial correlation with extent of diffusion, controlling for quality
A. N honorific awards (1961)	.34	.26
B. Prestige of highest award (1961)	.32	.25
C. Rank of academic department	.24	.17

N = 157

NOTE: Extent of diffusion is measured by the total number of citations in 1961 to all the physicist's work. Quality is measured by the total number of citations in 1966 to work published prior to 1961.

The consequences of the Matthew Effect in this case depend upon a question we have not yet considered: who are the early and late citers of the work under consideration? The papers generally followed a pattern of increasing diffusion. Who are the men who used a paper in 1966 but did not use it in 1961? If, in fact, the scientists producing the best work in the subject area of the paper knew of the work in 1961 and used it, then the limitation of diffusion due to the author's rank is likely to have little effect on the advance of the field. If, on the other hand, a substantial number of top scientists do not learn of the work and do not make use of it soon after it is published, then the operation of the Matthew Effect in this case may be dysfunctional for scientific advance. These data point to the next step in the analysis of the communication system of science. The study of the structural bases of awareness in science showed that knowledge flowed smoothly throughout the social system of physics; good work was known about equally to physicists in all parts of the system (see chapter 6). We must now investigate changes in patterns of utilization over time. Who are the men who will make immediate use of scientific work destined to be heavily utilized? What kinds of scientists use a discovery only after it has been recognized by others? In short, what roles are played by different types of scientists in the life history of a discovery?

A first step toward answering some of these questions has been taken. We used the sample of papers that were published in the 1963 volume of the *Physical Review* and subsequently received six or more citations in the 1966 *SCI*. The authors of these papers were then divided in terms of the quality of their lifework. Three groups were assembled: those whose work received 100 or more citations, 20 to 99 citations, and fewer than 20 citations. Finally, we computed the mean number of citations to the citers of these papers at two points in time—1964 and 1966. This process enables us to compare the citers of papers one year after they are published and three years after publication. The data are presented in table 6.

Table 6. Comparison of citers utilizing immediately and later papers published in the 1963 *Physical Review* by physicists of differing accomplishments

Number of citations to lifework of cited authors *	Mean number of citations to work of citers			
	1964	Number of citers	1966	Number of citers
100 or more citations	52	(27)	33	(26)
99–20 citations	37	(21)	38	(25)
19–0 citations	23	(25)	25	(40)

* As measured by number of citations in 1964 to their lifework.

If we concentrate only on citers of authors with less than 100 citations to their lifework, the results are clear. There are no changes in the average number of citations to citers of the work at time 1 and time 2. The average quality of research by citers of this work remains constant. Only when we note the utilization patterns of work by authors who have been cited more than 100 times in a single year does a significant difference appear. There is a substantial drop over time in the average number of citations to work by citers of this most eminent group of physicists. In 1964, only one year after publication, the citers of papers by these authors received, on the average, 52 citations to their own research. After two more years, the average quality of work by citers of the same work dropped to 33.

These data suggest a new temporal dimension to the utilization of research publications in physics. For most work there are no differences in the types of citers of work over time, at least for the quality of research produced by the citers. However, the earliest citers of authors whose research has received more than 100 citations appear to include, when looked at as a group, a higher proportion of scientific "influentials" than do the later citers of the same work. These earliest citers are those who probably are closest to the center of the communication system; who find out about relevant research quickly; and who incorporate the new knowledge into their own research rapidly. Often, these scientists hear about the results of research before they are put into print. Consequently, they have a head start on the use of the discovery. One should not underestimate the importance of this head start for the development of new ideas. After a short period of time—it seems to take only a couple of years—other scientists who produce work of lesser impact begin to use this published research. After only three years, the results of the discovery are fairly widely diffused throughout the social system of science, and the proportion of scientists that produce extremely high-quality research that cite the new work correspondingly declines. These data do not necessarily imply that after a few years the most distinguished scientists no longer use the discovery. They simply indicate that the most distinguished citers represent a smaller proportion of the total group of citers. A simple test of the variance of the quality of citers' research over time could be used to test the homogeneity of the groups of citers. An extension of this analysis of the composition of citers, looking at prepublication citers, might disclose even greater differentials in the types of men who utilize scientific research from one period to another.

THE RETROACTIVE EFFECT

So far we have analyzed the importance of the Matthew Effect in the early recognition of discoveries and in their wide diffusion. The Matthew Effect may also serve to focus attention retroactively on work of men who go on to become eminent.[22] To examine this aspect of the Matthew Effect we use the study of citations to the work of the 1,308 university physicists. We took all men who were thirty-five or younger in 1961 and counted the number of citations that

22. Merton points this out in the situation of retroactive recognition of the role of a junior scientist in collaboration: "Should the younger scientist move ahead to do autonomous and significant work, this work retroactively affects the appraisals of his role in earlier collaboration" ("The Matthew Effect," p. 58).

their pre-1961 work received in 1961 and in 1966.[23] We also counted the number of citations in the 1966 index to work published between 1962 and 1966. We then reversed the procedure of the earlier analysis, in which we controlled for citations at time 2 and looked for variations in citation at time 1. Here we controlled for assessed quality at time 1 and looked for variation in citation at time 2. The independent, or stratification, variable is the number of citations received in 1966 by work published between 1962 and 1966. This procedure basically allows us to control for assessed quality of work at time 1, split up our sample into those who became more successful after time 1 and those who did not, and finally look at the assessment of the quality of early work at time 2. The data indicate that work which is originally cited an equal number of times in 1961 is not cited equally in 1966. The men who went on to publish important work after 1961 experience sharp increases in citation to their earlier work. On the other hand, those scientists who do not go on to publish important work see a decline in the number of citations to their early work.[24] The zero-order correlation between citations in 1966 to pre-1961 work and citations in 1966 to recent work (1962–66) is .47. When we "control" for the number of cita-

23. We limited our study to young scientists because we did not think that continued success would bring about a retroactive interest in the work of someone who was already quite well known at the time of publication of his early work. If such an increase were found, it would be difficult to attribute it to the retroactive aspect of the Matthew Effect.

24. This conclusion is, of course, not based on the correlation or partial correlation statistics since these tell us nothing about specification or interaction effects. The conclusion is based upon the tabular treatment of the data presented here. Mean Number of Citations in 1966 to Pre-1961 Work by the Number of Citations the Same Work Received in 1961 and the Number of Citations in 1966 to the Author's Work Published Between 1961 and 1966 (University Physicist Sample—35 Years of Age or Less)

	Number of Citations in 1966 to Work Published 1961–66	Mean Number of Citations to Pre-1965 Work in: 1961	1966		Difference	Number of Men
Heavy early citations	50 or more	26	46	=	+20	20
(10 or more)	49–10	24	24	=	0	38
	9–0	21	8	=	−13	29
Light early citations	50 or more	3	27	=	+24	11
(0–9)	49–10	2	12	=	+10	120

NOTE: The correlations reported in the text are based on an N of 598.

tions the early work received in 1961, we get a partial correlation of .25.

One possible interpretation of these data is that the scientific community is merely making ceremonial citations to the early work of colleagues of new prominence. This seems unlikely, for ceremonial citations might be better made to the more recent work of these men, the work which has elevated their reputation. It is more likely that the audience is actually going back and reexamining the early work of men who have more recently produced outstanding and recognized contributions. It is quite possible that a physicist's later work has made his earlier work more relevant. Often scientists refer in their papers to earlier work. This earlier work may, from a purely substantive point of view, become more significant as it is developed. If this were true, then the increased rate of citations to early work may not be a result of the current location of the author in the stratification system but be a result of a real increase in the "quality" of the work. This leads to the perhaps obvious conclusion that a judgment of the quality of scientific work is time-bound. Work can improve or deteriorate with age.

RESISTANCES OR DELAYED RECOGNITION

We have been examining the impact of stratification on the diffusion of scientific ideas. There is one special case of delayed diffusion that merits our consideration: resistance to scientific discoveries. Historians and sociologists of science have frequently noted that important discoveries are sometimes ignored and at other times actively resisted.[25] Unfortunately, it has been difficult to collect the kinds of data necessary for systematic study of this phenomenon. The data on citation patterns over time can be used as the first step toward such a systematic study. We can operationally define resistance—or, as we prefer to call the phenomenon, delayed recognition[26]—as those cases in which papers published prior to

25. R. H. Murray, *Science and Scientists in the Nineteenth Century* (London: Sheldon, 1925); Bernard Barber, "Resistance by Scientists to Scientific Discovery," *Science* 134 (September, 1961):596–602 (reprinted in Barber and Hirsch, *The Sociology of Science;* all quotations are from the reprint); Robert K. Merton, "Resistance to the Systematic Study of Multiple Discoveries in Science," *European Journal of Sociology* 4 (1963):237–82.

26. Delayed recognition would seem to be a more inclusive term than resistance. The latter implies a conscious rejection by a scientific audience. Many of the cases given of resistance do not involve any conscious rejection but rather a simple ignoring of work. Delayed recognition describes a phenomenon of which resistance is a special case.

1961 received a substantial number of citations in 1966 and few or no citations in 1961. This procedure is, of course, dictated not by the requirements of ideal design but by the availability of data. Ideally there should be a longer period of time between the two measurements of impact. However, we must also note that five years now probably sees more science come and go than fifty years did in the nineteenth century—the favorite hunting ground for collectors of cases of resistance.

Noting, then, that the design is less than ideal, let us estimate the frequency of delayed recognition and analyze some of its sociological sources. In order to identify cases of delayed recognition, we must limit ourselves to work which is indeed recognized at time 2. The sample, therefore, consists of 10 percent of the papers from all the fields of science (in the *SCI*) published prior to 1961 and receiving ten to twenty-nine citations in 1966, and all those papers published prior to 1961 which received thirty or more citations in 1966. This sampling procedure yielded a list of 587 papers. Of these 587 papers, 74, or 13 percent, received three or fewer citations in 1961. These papers that received at least ten citations in 1966 and three or fewer in 1961 shall be considered to have received delayed recognition. First, let us note that delayed recognition is relatively rare. On the basis of our sample we estimate that there are only 380 papers in the scientific literature published prior to 1961 that received ten or more citations in 1966 and three or fewer in 1961. Apparently the evaluation and communication system of institutionalized science operates so that only a relatively small number of papers that later turn out to be significant are overlooked at the time of their publication. Despite the rarity of the occurrence, we still wanted to investigate the sociological sources of delayed recognition, as information on deviant cases sometimes also provides knowledge on normal occurrences.

Bernard Barber suggests several possible conditions making for delayed recognition.[27] He begins his analysis by showing that new concepts and methodologies which are opposed to existing scientific ideas are often resisted. This type of resistance may not be entirely contradictory with science's self-ideal. Skepticism about new ideas until they have been fully developed and adequately demonstrated falls well within the approved value system of scientists. Also, delayed recognition of ideas truly ahead of their time does not imply

27. Barber, "Resistance to Scientific Discovery."

any sociological influences on science's development. In order to demonstrate the existence of an element of "irrationality" in the evaluation system of science, it is necessary to show that "resistance" is not randomly distributed, that the work of low-ranking men is disproportionately overlooked. This is, in fact, one of Barber's major hypotheses and of direct relevance for our major theme: the significance of professional standing for the reception of a scientist's work.

Barber suggests that "sometimes, when discoveries are made by scientists of lower standing, they are resisted by scientists of higher standing partly because of the authority the higher position provides."[28] He then goes on to give examples of work by young and little-known scientists that were ignored. Among these were the work of the mathematician Niels Abel, the mathematician Ohm, and the geneticist Mendel. He also suggests that sometimes "men of higher professional standing sit in judgment on lesser figures before publication and prevent a discovery's getting into print."[29]

Barber's paper was a major contribution because it brought to our attention a problem that has itself experienced delayed recognition. However, our research leads us to question the relevance for modern science of this type of delayed recognition. Indeed, it is possible that the sociology of science may be hampered by an overdependence on historical material. In the study of an area of human activity that has grown and progressed as rapidly as science and has experienced such rapid institutionalization in the twentieth century, we may misallocate our research effort by assuming that phenomena of past significance are still important. Let us first consider the possibility of a decent piece of science going unpublished. We know from the work of Zuckerman and Merton that most journals in the hard sciences publish a majority of the papers submitted to them.[30] The *Physical Review* publishes approximately 80 percent of submitted papers. The editors of these journals tell us that rejections are limited predominantly to work that is not "plausible." When we couple these high acceptance rates with the knowledge that most papers published by the best journals are of relatively low quality and the fact that there are so many scientific journals, we may conclude that the chances of any truly valuable scientific work

28. Ibid., p. 550.
29. Ibid., p. 552.
30. Zuckerman and Merton, "Patterns of Evaluation."

being refused publication are so slight as to make the problem of limited sociological relevance.

Now let us consider what we might call the "Mendel case," where a significant discovery is published and then ignored, to some extent as a result of the author not having a high status in

Table 7. Distribution of authors of delayed recognition papers and others (sample of papers having 10 or more citations in 1966: all fields included in *SCI*)

Characteristics of authors	Delayed recognition papers *	Other papers
A. *Age*		
40 or under	14%	15%
41–50	38	45
51–60	26	22
Over 60	22	17
	100%	99%
	(58)	(419)
B. *Institutional location*		
Academic departments (distinguished and strong)	43%	48%
Academic departments (all others)	19	21
Nonacademic	28	28
Retired	10	3
	100%	100%
	(58)	(419)
C. *Number of honorific awards*		
0	36%	41%
1 or 2	24	23
3 or more	40	37
	100%	101%
	(58)	(419)
D. *Scientific repute* (N *citations to work in 1961*)		
100 or more	16%	27%
99–20	32	49
19–0	52	24
	100%	100%
	(74)	(513)

* Delayed recognition papers are those which received 10 or more citations in 1966 and 3 or fewer citations in 1961. Sections A-C exclude cases for which AMS information was unavailable.

science.[31] We suggest that this case is almost as unlikely to occur as the case of nonpublication. First, we would be unlikely to find a contemporary Mendel working in an obscure monastery; modern day Mendels would be in a university science department, or a government or industrial laboratory. Also we know that the evaluation system of science operates so efficiently that most Mendels would be in the top university departments (see chapter 4). We would suggest that modern science gives such great indications of universalism and rationality that the only cases today of important discoveries going unrecognized for more than a few years would be those cases of delayed recognition for truly intellectual reasons—that is, the discoveries that are truly ahead of their time.

We hypothesized that delayed recognition in modern science was either solely the result of the content of the discovery or partly influenced by differentials in visibility of the author that would lead to short-term delays in the recognition of a minority of important discoveries. The data are presented in table 7.[32] They suggest that the content of papers is probably more important than the social characteristics of their authors in bringing on delayed recognition. Papers receiving delayed recognition are no more likely to be written by young men than papers receiving immediate recognition. Likewise, institutional location and number of honorific awards both fail to distinguish the authors of delayed recognition papers from those not experiencing delayed recognition. The only variable which did make a significant difference was the number of citations to the author's other work at the time of publication. This finding is consistent with those presented above and once again emphasizes our conclusion that the more one's past work is being used, the higher the probability of new work being quickly recognized and diffused.

CONCLUSIONS

Let us briefly summarize the major conclusions of this set of investigations. Perhaps most important, we have shown that modern physical and biological science does approach its ideal of universalism in the reception of scientific discoveries. All the data indicate that the assessed quality of papers at time 2 is a far more important

31. As Barber points out, Mendel was resisted because of his low professional standing and also because his ideas were considerably ahead of his time.

32. The data in this table are used for descriptive purposes and are therefore percentaged in the direction of the dependent variable.

determinant of the paper's initial reception at time 1 than the author's rank in the stratification system at time 1; significant work tends to be utilized regardless of who has produced it. We would suggest that only a small fraction of significant work is overlooked for more than a few years. Papers which do experience delayed recognition most often remain unknown because of their content and not because of the author's location in the stratification system.

The Matthew Effect, as we have used the concept, stands for the influence of all aspects of stratification on the reception of scientific ideas. When the quality of work is controlled, the Matthew Effect can be seen to have a greater influence on the extent of diffusion of a scientist's complete work than on any particular paper. Good papers have a high probability of being recognized regardless of who their authors are; but lesser papers written by high-ranking scientists are more likely to be widely diffused early than are lesser papers by low-ranking authors. The Matthew Effect also serves to focus attention on the work of little-known men who collaborate with high-repute scientists, and possibly to increase retroactively the visibility of the early work of scientists who go on to greater fame. The generalization of the concept of the Matthew Effect has led us to raise questions about its functional consequences for scientific advance. When the Matthew Effect is applied to all discoveries of equal quality, we can see that it may result in the temporary ignoring of some significant discoveries. This research points to the need of a more complete exploration of not only the conditions under which the Matthew Effect operates but also the conditions under which it is functional, dysfunctional, or nonfunctional for scientific advance.

Finally, we suggest that there is possibly greater sociological discontinuity between little science and big science than we have assumed in the past. The sociology of science has developed predominantly as an offshoot of the history of science. The leaders of the specialty were either trained as historians or began their sociological investigation with historical topics. It is natural that latecomers to the field have been influenced by problems and perspectives which emerged from historical analysis. Recent research, however, suggests that the social organization of science has changed so drastically in this century that there may be real discontinuities between what science as a social institution is today and what it has been in the past. It is possible that resistance to scientific discovery is not a significant problem in contemporary science. Our purpose is not to suggest that sociological investigation of the historical development of science is of limited value. On the contrary, we would like to suggest that historical investigation is of great value if put in proper perspective. To take only one example, there is certainly an overwhelming need for sophisticated studies of the processes of institutionalization. We are suggesting that when studying the current social organization of science we may be misled by use of historical examples or the assumption that past problems are also current ones.

8 THE ORTEGA HYPOTHESIS

A key proposition of the functional theory of stratification is that those people who perform the most highly valued functions in society receive the most rewards. In science we have shown that the most eminent scientists have produced high-quality work. Contributing to scientific progress is, of course, the most highly valued activity in science. The question that we have not considered in detail is the extent to which the eminent scientists who produce the most important work are dependent upon the work of the noneminent scientists. When Newton said with typical scientific humility, "If I have seen farther, it is by standing on the shoulders of giants," did he mean quite literally that his work was dependent solely on the great scientists who preceded him, or were those giants in turn standing on a mountain of dwarfs?

If the work of eminent scientists could not have been produced without the contributions of their noneminent colleagues, then perhaps the average scientist is being deprived of his due share of recognition. In this chapter we present data bearing on the question of whether scientific progress is built on the labor of all "social classes" or is primarily dependent on the work of an "elite." In the past, historians and philosophers of science have attributed much of the growth of science to the work of the average scientist who, it is suggested, has paved the way with his "small" discoveries for the men of genius—the great discoverers. This hypothesis is asserted in many sources, but perhaps no more clearly than in the words by Jose Ortega y Gasset:

> For it is necessary to insist upon this extraordinary but undeniable fact: experimental science has progressed thanks in great part to the work of men astoundingly mediocre, and even less than mediocre. That is to say, modern science, the root and symbol of our actual civilization, finds a place for the intellectually commonplace man and allows him to work therein with success. In this way the majority of scientists help the general

> advance of science while shut up in the narrow cell of their laboratory, like the bee in the cell of its hive, or the turnspit of its wheel.[1]

Ortega seems to be suggesting that average scientists, working on relatively unambitious projects, make minor contributions, and that without these minor discoveries by a mass of scientists the breakthroughs of the truly inspired scientists would not be possible. Thus the work of the great scientist is built upon a pyramid of small discoveries made by average scientists. This view of science is widespread. Some even go so far as to maintain that scientific advance is more dependent upon the small discoveries of the many average scientists than the breakthroughs of the great scientists. Lord Florey, a recent president of the Royal Society, expressed this point of view:

> Science is rarely advanced by what is known in current jargon as a "breakthrough," rather does our increasing knowledge depend on the activity of thousands of our colleagues throughout the world who add small points to what will eventually become a splendid picture much in the same way the Pointillistes built up their extremely beautiful canvasses.[2]

There are, of course, a number of assumptions in this view of science. Consider two: first, it is assumed that the ideas of the average scientist are both visible and used by the outstanding scientist; second, it is assumed that the minor work is necessary for the production of major contributions. In short, it is proposed that the work of the average scientist is indispensable if science is to advance. Little empirical evidence exists to substantiate these widely held beliefs. We shall examine data bearing upon the validity of this view of scientific progress. In order to make an empirical test of this conception manageable, we confine ourselves to one of its several aspects and to only one field of science. We shall examine the work of several samples of physicists and analyze what work these men built upon in making their discoveries.

We do not intend to suggest that great discoveries in science by an Einstein or a Lee and Yang are not preceded by numerous "smaller" discoveries, or that great discoveries do not in turn

1. J. Ortega y Gasset, *The Revolt of the Masses* (New York: W. W. Norton, 1932), pp. 84–85.

2. Quoted in J. G. Crowther, *Science and Modern Society* (New York: Shocken Books, 1968), p. 363.

stimulate a multitude of lesser ones.[3] We will suggest that even the scientists who make these "smaller" discoveries come principally from the top strata of the scientific community. In the proper perspective of the history of science, "normal science" as Kuhn refers to it, is not done by the average scientist but by the elite scientists.[4] Indeed, in the longer perspective, the work of many of today's outstanding scientists, even the work of such scientists as Nobel laureates and National Academy members, may turn out to be a minor footnote in the history of science.

The question that we consider is, how many scientists are contributing *through their published research* to the movement of science, and how many are not? There are, of course, many other ways to contribute to scientific progress than through published research. The scientists who are primarily teachers, administrators, or technicians may play crucial roles in scientific development. We do not intend to downgrade the importance of these roles. Nevertheless, it is still a valid question to ask how many scientists contribute to scientific progress through their published work, if we keep in mind that to the list of contributors of this type we must add the names of contributors of other types.

As we noted in chapter 3, Derek Price, following Lotka, has estimated that the number of scientists producing "n" papers is approximately proportional to $1/n^2$. This inverse-square law of productivity estimates that for every 100 authors producing one scientific paper, there are only 25 who produce two, 11 who produce three and so on.[5] Using Price's model, we can estimate that roughly 50 percent of all scientific papers are produced by approximately 10 percent of the scientists. What remains problematic is the extent to which the 10 percent of the scientists who produce 50 percent

3. For detailed and informative discussions of fluctuations in rates of discoveries in the history of science, see P. A. Sorokin and R. K. Merton, "The Course of Arabian Intellectual Development, 700–1300 A.D. A Study in Method," *Isis* 22 (1935):516; P. A. Sorokin, *Social and Cultural Dynamics* (New York: The Bedminister Press, 1962), vol. 2; J. Ben-David, "Scientific Productivity and Academic Organization in Nineteenth-Century Medicine," *American Sociological Review* 25 (1960):828. For a qualitative treatment of the same idea, see G. Sarton, *History of Science and the New Humanism* (New York: Henry Holt, 1931), especially pp. 34–42.

4. T. S. Kuhn, *The Structure of Scientific Revolutions*. We use the term "elite" here and throughout in a statistical sense to refer to the small group of eminent scientists who publish the most, are most frequently cited, and occupy the most prestigious positions. In fact, Zuckerman notes that this statistical elite does form a fairly cohesive social group. See H. Zuckerman, "Stratification in Science."

5. Price, *Little Science, Big Science*.

of the research publications are dependent on the other 90 percent of research scientists and the 50 percent of the total research they produce. If the bulk of the scientific community produces work that is rarely used, that is, infrequently cited in the work of outstanding scientists, then this may indicate that the work does not materially advance the development of science. The basic question to ask is, what are the intellectual sources of influence on the production of scientific research of varying quality? If Ortega is correct, the work of scientific frontiersmen will to some extent be dependent upon the work of the vast majority of physicists.

CITATION PRACTICES OF ACADEMIC PHYSICISTS

We collected data which illustrate the citation practices of academic physicists. One set of data consists of the citations made by 84 university physicists in the paper by each most heavily cited in the 1965 *SCI*. We consider this to be the physicist's outstanding piece of work as gauged in 1965.[6] The 84 physicists are in fact, a subsample of a sample of 120 university physicists (see Appendix A). A second set of data consists of information on a one-third random sample of the scientists who were cited in the best paper of each of the 84 physicists. For the sample of 385 cited authors we collected data that enabled us to locate them in the stratification system.[7]

6. Since we were primarily interested in the influences upon the pure research of these scientists, and were using citations to measure that influence, we decided to omit from consideration any paper that was a "review article." This was done because we were not interested in a review of the literature and felt that citations to an enormous amount of literature of this type would distort our results. In such cases where a review article was the most heavily cited, we took the next most heavily cited paper and included it in our sample. In looking up the cited authors within these individual papers, we were faced with the problem of what to do with collaborating authors. We decided to treat a paper as a single unit and locate information on all collaborators in the research team. To some extent, junior collaborators who often were students of the senior authors dropped out of our sample because no information could be found for them in *American Men of Science*. But it can be seen that the average level of eminence of cited authors is weighted against our hypothesis to some extent since our sample included junior men of distinctly less eminence than their senior collaborators. The subsample excludes all totally nonproductive scientists, those whose work had received absolutely no citations, and those who produced only review articles that were cited.

7. One of the limiting features of this study has to do with the collection of information on the scientists who are being studied. Perhaps the best source for independent information short of a questionnaire sent to a sample of physicists is the *American Men of Science*. The sample only includes men who were listed in *AMS*. Only about one-half of all cited authors appeared in these volumes. We wanted to see whether there were systematic differences in the types of scientists who appear

A basic assumption in this analysis is that the research that scientists cite in their own papers represents a roughly valid indicator of influence on their work. Of course, not all citations represent direct and specific influence. Everyone knows of cases in which scientists ceremonially cite friends, colleagues, mentors, or eminent people in the field. Sometimes a citation to an expert in the field serves the function of legitimating the new paper. Even when we cite work which has influenced us, it is difficult for the reader to know when a cited work represents a significant, even necessary, antecedent to our work, as opposed to a tangentially relevant piece of work, in which we are merely demonstrating our "knowledge of the literature." Furthermore, relevant and influential material is passed from one scientist to another through private communications which, though often mentioned in today's age of big science, sometimes do not show up as citations. However, a reasonable case can be made that citations generally represent an authentic indicator of influence.[8]

Let us consider the process through which we decide what to cite in our papers. Some part of our citations will be clearcut. We will, of course, cite papers that contributed directly to the current state of knowledge in our problem area. In this chapter, for example, such a citation would be to the work of Derek de Solla Price. There would be, however, another group of references that would be more questionable. These would be references to people who have done work in the area but have not had a direct influence on the paper. Why do we cite some of these rather than others? We tend to cite those scientists who have the highest visibility. Originally scientists gain visibility by publishing significant research. After such visi-

---

in *AMS* and those who do not. We found, of course, that a large proportion of the men who were not found in *AMS* were foreign scientists. The second largest group of individuals who could not be found turned out to be students at the institution where the research was being done. It is not infrequent that older, more eminent scientists collaborate with their doctoral students. It should be added that *AMS* tends to include more academic scientists than scientists located in industrial concerns. We did find that the average number of citations to men in *AMS* was approximately 1.5 times that of those not found in it. We also found, and this was to be expected, that the cited authors in the best work were more often found in *AMS* than those cited in lesser-quality work. To the extent that the *AMS* does not include the less eminent members of the scientific community, our sample of cited authors overrepresents eminent scientists.

8. The extent to which unpublished work is being cited in leading journals is increasing rapidly, at least in physics. Second to articles published in the *Physical Review*, private communications are the most cited source of information in contemporary physics.

bility is gained, they enjoy a halo effect as their research gains additional attention due to their visibility. Thus if we consider the sum of a scientist's citations, some part will be due to the halo effect. But the size of the halo effect will probably be directly related to the significance of the scientist's research. The processes of objective evaluation of contributions and the subjective working of the halo effect, work together to create substantial gaps between the number of citations received by members of the elite and the average scientist.

The halo effect would cause us to cite a scientist whose work was not directly influential. But we are primarily interested in situations in which work that is directly influential is not cited. The norms of science require scientists to cite the work that they have found useful in pursuing their own research, and for the most part they abide by these norms. Moreover, the audience of the work generally takes citations as an indicator of influence. We only have to think of the number of times we have taken a quick glance at the acknowledgments and references in books and papers with the intent of noting the influence on a piece of work, to realize that at the very least citations do indicate intellectual connections. Sometimes, however, a crucial intellectual forebear to a paper is not cited. This is rarely due to direct malice on the part of the author but more often to oversight or lack of awareness. This most frequently occurs in cases where a scientist's work has had such a deep impact on the field that the ideas have become part of the accepted paradigm, and explicit citation is not deemed necessary. Only the work of a handful of scientists ever achieves this status, and they generally receive very heavy citation anyway. (The work of Einstein, for example, was cited 281 times in the 1970 edition of *SCI*.) In the cases of omitted citations to less influential work, we can assume that such omissions are random in nature and, although we may fail to cite the important work of a particular scientist, you will not make the same error. In general, the procedure of using citations as an indicator of influence probably errs on the side of over-inclusion rather than exclusion of significant influences.

CHARACTERISTICS OF CITED AUTHORS

The characteristics of the sample of 385 authors cited in the best papers of 84 university physicists are presented in table 1. We want to compare the characteristics of these cited authors with those of the population of physicists. In many cases, this is difficult because

Table 1. Marginal distributions of the social and individual characteristics of the authors cited in 84 papers, with comparative figures for the entire field of physics

Social and individual characteristics of cited authors			Comparative "population" statistics	
Current affiliation				
University	72%		43% (University and College or nonacademic research laboratories combined)	
College or nonacademic research laboratories	10			
Industry	10		34	
Government	8		11	
	100%		88%	
		(385)		(26,698) *
Rank of department of those in academic departments				
Distinguished (top nine)	60%		21%	
Strong and good	23		42	
Lesser universities and colleges	17		37	
	100%		100%	
		(299)		(1,308)†
Number of honorific awards				
0	32%		73%	
1	18		15	
2–3	23		9	
4 or more	27		3	
	100%		100%	
		(385)		(1,308) †
Quality of scientific output (number of citations: 1965 SCI)				
Under 15	25%		67%	
15–59	33		25	
60 or more	43		8	
	101%		100%	
		(385)		(1,308) †

*Source: American Science Manpower 1964: A Report of the National Register of Scientific and Technical Personnel, National Science Foundation, NSF 66–29.

†These figures are drawn from the sample of 1,308 university physicists.

the population parameters are unknown. We have, therefore, used the sample of 1,308 university physicists as the comparison group.[9] Of course, this sample of 1,308 is itself an elite group and far from representative of the more than 25,000 American physicists. As the data of table 1 indicate, physicists in the top strata are far more

9. For a full description of this data set, see Appendix A.

likely to be cited than those below the top. Whereas 73 percent of the sample of 1,308 university physicists had no awards listed after their names in AMS, only 33 percent of the cited authors had no awards.[10] The same results are found when we examine citations to the work of the cited authors and the university sample. On the average, the cited authors received 119 citations to their life's work in 1965, while all the authors listed in the 1965 *SCI* received a mean of six citations. Further, although only 8 percent of the 1,308 university physicists averaged 60 or more citations, 43 percent of the cited authors exceeded this lofty number.[11]

The figures presented in table 1 lead to the conclusion that most of the work used by university physicists in their best papers is produced by only a small proportion of those who are active in the field. It is equally important, however, to note that a significant *minority* of cited work is being produced by nonelite physicists. So far, we have not made any distinctions among the citing papers. We have just considered the references in the "best" papers of a stratified random sample of university physicists. Many of these "best" papers may have been of relatively little significance. If the Ortega view of science is correct, we should find the top papers making just as much use of the work of little-known physicists as the less significant papers. We shall present three sets of data to test this hypothesis. Obviously, the number of citations to these "best" papers varied greatly. Some papers received only one or two citations; others received over twenty or thirty. We shall now see the extent to which authors of papers of varying quality depend upon the work of elites and nonelites.

As the data of table 2 indicate, highly cited papers, more often than those receiving few citations, make use of high-quality work produced predominantly at the nine most distinguished departments.[12] We see that a mere 7 percent of the citations in the most

10. It would probably be safe to assume that more than 90 percent of the population of physicists have no awards.

11. Inclusion in the scientific elite could be a function of longevity if the bulk of citations went to older scientists. The data do not support this possibility. The pattern of citations by scientists in various age groups suggests that older scientists tend to cite work by older scientists; younger scientists tend to cite most often the work of other young scientists. Over 50 percent of the cited authors, however, were under fifty years old. The cited authors slightly overrepresented older scientists because the sample of 84 citers overrepresented older scientists.

12. For the prestige ranking of departments see Cartter, *Quality in Graduate Education*. The data presented in table 2 are in fact the same data that were presented in table 8 of chapter 6. Here we present the data in tabular form and examine them from a slightly different perspective.

Table 2. The distribution of citations in individual papers of varying quality according to the characteristics of the cited scientists

Characteristics of cited scientists	Quality of citer's "best" paper: High (20 or more citations)	Medium (10–19 citations)	Low (0–9 citations)
Rank of department			
Distinguished	60	50	36
Strong-good	14	19	19
Lesser	7	12	18
No academic affiliation	19	19	27
	100% (95)	100% (139)	100% (151)
Quality of scientific output (number of citations)			
High (60 or more)	54	48	33
Medium (15–59)	28	30	36
Low (less than 15)	18	22	31
	100% (95)	100% (139)	100% (151)
Prestige of highest award			
Nobel Prize, National Academy member	45	32	25
Other honorific awards	15	8	12
Only fellowship plus no awards *	40	60	63
	100% (95)	100% (139)	100% (151)

* Fellowships such as the Guggenheim, Sloan, Rockefeller, and Fulbright were here considered as honorific awards as distinct from other post-doctoral fellowships.

highly cited discoveries go to scientists working in the lower-prestige university departments and colleges, while 60 percent are to scientists at distinguished departments. Even in the papers receiving less than 10 citations, the work of those at top universities is cited considerably more frequently than those working in lower-prestige departments. The best papers predominantly cite other significant papers. Fifty-four percent of the citations in papers receiving 20 or more citations, and 33 percent in those receiving less than 10 citations, go to the work of scientists who have received 60 or more citations. Finally, take the extent to which high-quality papers rely on the work of Nobel laureates and National Academy members. Forty-five percent of the citations in these papers go to the work of no more than 200 scientists and their collaborators. Although the lower-quality papers do not cite these "elites" to the same extent, the work of the elites receives proportionately greater use in these papers as well.

Throughout this chapter, we have defined important discoveries simply by the number of citations that they have received. As a further test of the Ortega hypothesis, we asked a well-known physicist to list the five most important contributions to elementary-particle physics in the last ten years. Of course, in many ways this procedure falls short of the rigorous study needed to further test the hypothesis. It would be useful, for example, to have a broad, stratified panel of judges evaluate the merits of various pieces of research, and then to look at the citation patterns in papers judged to be of highest impact. It is noteworthy, however, that the five papers chosen by our informant received a mean of 67 citations in the 1965 *SCI*.[13] We took all the journal and private communication citations in these five papers, and then located the cited authors in the stratification system of physics. The five papers cited a total of 51 articles (not counting self-citations) involving 126 authors, 19 of whom were located at foreign universities and foreign research laboratories.

The data corroborate the earlier findings. Of the 107 American scientists cited in these five "pathfinding" papers, *all but one* were located at one of the top nine physics departments in the United States or at such distinguished laboratories as Brookhaven or the Lawrence Radiation Laboratory. All 51 articles were produced at one of these top nine departments or laboratories. The average number of citations to the cited authors is equally impressive. This group had a mean number of citations to all their work of 69 in the 1965 *SCI;* 74 percent of these authors had more than 60 citations to their work in 1965. Among the 107 cited authors were a number of younger and not yet widely recognized scientists who were co-authors of more eminent colleagues. The mean number of citations to either single authors or the most highly cited author of collaborative papers is 134.[14]

ADDITIONAL TEST OF THE ORTEGA HYPOTHESIS

Since this type of subjective sampling procedure may indeed be methodologically suspect, we decided to perform one final test of the

13. These five articles included, for example, Lee and Yang's now famous paper on parity conservation. Three of the authors turned out to be Nobel Prize winners; the others, members of the National Academy of Science.

14. Nobel laureates in physics who received their prize between 1950 and 1964 averaged 130 citations to their lifework in the 1965 *SCI*.

Ortega hypothesis. We replicated the essential aspects of the study design, using a set of independent data. We had a complete list of all papers that were cited three or more times in the *Physical Review* in 1965.[15] This list contained more than 3,000 scientific articles and substantive letters. A few of these papers were cited often; most received less than five citations. Since we are primarily concerned with the pattern of citation in influential papers, we initially examined the ten papers that were most often cited in the *Physical Review*. After identifying these "super" papers, we listed the scientific articles that the authors of these influential papers cited. Finally, we counted the number of citations received in 1965 by authors of papers cited in the "super" articles. This procedure can be clarified by reference to a specific case. Murray Gell-Mann produced the most heavily cited article on the list. It received a total of 49 citations in the *Physical Review* in 1965. We took Gell-Mann's paper and listed the references in it. A total of 33 publications, or 55 scientists, were cited in the paper. We then noted the number of citations that the lifework of each of these 55 scientists had received in the 1965 *SCI*. This process was followed for all of the scientists cited in the ten "super" papers. Thus while we are examining only these ten most highly cited papers, we studied a total of 299 authors cited in the ten papers.

The results obtained from this replication offer further evidence in support of the earlier findings. It turns out that authors cited in these ten papers were scientists who, on the average, had produced truly outstanding scientific work. In 1965 the 299 cited scientists produced research that received an average of 135 citations. Since this figure includes citations received by beginning scientists yet to make their mark, who are collaborating with their more eminent colleagues, the statistic is actually lower than it would be otherwise. In fact, if we take only single-author papers and the most cited author in each collaborative paper, and compute the average number of citations to the author's lifework, the mean is increased to 175 citations. Clearly these data lend added weight to the counter-hypothesis that work which is used by the producers of outstanding research is itself produced by a small minority of scientists. The work of the average researcher is rarely the work that is influential in the production of high-impact scientific research.

A question remains to be answered. What is the quality of re-

15. We thank Dr. Cullen Inman of the American Institute of Physics for making these data available to us.

search that is cited in the work of the physicists whose papers receive fewer citations than those of the ten "super" papers? Using the same *Physical Review* list, we drew a small random sample of papers that had received from 23 to 3 citations in 1965. Papers that received 23 citations were of approximately the same impact as some of the top ten, since the range of citations to the "super" papers was from 49 to 24 citations. This small sample consisted of 36 papers. Within these papers, references were made to 492 communications. We computed the number of citations in 1965 to the 837 physicists who authored these 492 papers. We examined citation rates to cited authors who produced single-author papers and those in collaborative teams whose work was most often cited. The data are presented in table 3. The table suggests that even authors

Table 3. Citation patterns of papers cited in the 1965 *Physical Review*

Number of citations to the paper in the 1965 *Physical Review*	Mean number of citations in 1965 to the lifework of major authors* cited in the papers
Top ten papers (24–29 citations)	175 (174)
20–23 citations	169 (88)
10–19 citations	158 (215)
5–9 citations	149 (124)
3–4 citations	85 (65)

* "Major author" was simply defined as all single authors and that author in a collaborative set that received the highest number of citations to his lifework in the 1965 *SCI*.

of less than super-quality papers were predominantly influenced by work of elites. While the top ten papers made use of work produced by physicists who received an average of 175 citations to their lifework, the average quality of cited work found in papers receiving 5 to 9 citations is not appreciably lower. Only when we examine the citation patterns in papers which received 3 or 4 citations is the average quality of work cited significantly lower. But even in this group, the scientists being cited are among the elite insofar as the quality of their work goes. These high averages are not due to a handful of extreme cases. Forty-one percent of the 837 cited physicists received more than 100 citations; another 13 percent from 60 to 99. Thus a total of 54 percent received more than 60 citations, a figure which is similar to those presented in table 2. Only eleven percent of the cited authors received less than 5 citations to their lifework, and 90 percent of the scientists comprising this 11 percent were coauthors on papers for which one of the other authors was more heavily cited. In short, there were virtually no cited authors whose work was not of above-average quality.

Consider once again a set of comparative statistics. First, about one-half of all papers that are published in the more than 2,100 source journals abstracted in the *SCI* do not receive a single citation during the year after the paper is published; second, that the average cited author in the 1965 *SCI* received a mean of 6.08 citations to his lifework. These data offer further support for the hypothesis that even the producers of research of limited impact depend predominantly on the work produced by a relatively small elite.

CONCLUSIONS

Let us consider, then, some general conclusions that may be drawn from the findings reported in this chapter. The data allow us to question the view stated by Ortega, Florey, and others that large numbers of average scientists contribute substantially to the advance of science *through their research*. It seems, rather, that a relatively small number of physicists produce work which becomes the base for future discoveries in physics. We have found that even papers of relatively minor significance used to a disproportionate degree the work of the eminent scientists. Although the conclusions of this chapter may be reasonably clear, the implications of these data for the structure of scientific activity, at least in physics, need careful consideration.

Consider only one problem emerging out of the findings: the size of the research establishment of modern science. If future research in other fields of science corroborate our results, what does this say about the relationship between the number of scientists and the rate of advance in science? Is it possible that the number of scientists could be reduced without affecting the rate of advance? The data would seem to suggest that most research is rarely cited by the bulk of the physics community, and even more sparingly cited by the most eminent scientists who produce the most significant discoveries. Most articles published in even the leading journals receive few citations. In a study of citations to articles published in the *Physical Review,* we found that 80 percent of all the articles published in 1963 were cited 4 or fewer times; 47 percent once or never in the 1966 *SCI.*[16] Clearly, most of the published work in even such an outstanding journal makes little impact on the development of science. The basic question emerges: could the same rate of advance in physics be maintained if the number of active research physicists were to be sharply reduced?[17]

Several criticisms of our position are possible.

1. The data indicate that about 15 to 20 percent of the work cited in significant discoveries is produced by "average" scientists. Could the important discoveries have been made using only the work of eminent scientists? It could be maintained that the 20 percent of references produced by, let us say, 80 percent of researchers are just as crucial for scientific advance as the 80 percent of references produced by 20 percent of researchers. To suggest an answer to this criticism we must make explicit an idea which is implicit in much work done in the sociology of science. Our entire analysis is dependent upon the assumption that no one scientist, elite or nonelite, is indispensable for scientific advance. The study of independent multiple discoveries leads to the conclusion that, if a particular scientist had not made a particular discovery, it is only a matter of time—probably a relatively short period of time—before the discovery will be made by another scientist.[18] The history of science offers many examples of discoveries made independently by two or more scientists within a short period of time. Merton has suggested that multiple discoveries, rather than being a rare occurrence, are the norm. Furthermore, many discoveries which are not multiples are forestalled multiples, as most scientists will stop working on a problem when they learn of the success of a competitor. As we learned from *The Double Helix,* if Watson and Crick had not made their historic breakthrough it probably would have been made in short order by Pauling. Most

16. See table 1, chapter 7.

17. We, of course, are not suggesting that people already in physics be dismissed from employment. We are suggesting that it might be possible to cut back the number of entrants to physics without reducing the rate of advance.

18. Merton, "Singletons and Multiples."

scientists working on important problems realize that there are many others working on the same problems. Indeed, chance often plays its part in determining who makes a discovery first.

If the work done by any scientist, elite or nonelite, can be replaced by work done by other scientists, how do we evaluate the extent to which a particular scientist is necessary for scientific advance? Merton defines the scientific genius as a man who is involved in multiple multiples—the functional equivalent of many other scientists.[19] Although no one citation or one man is crucial for any scientific discovery, the scientist who writes one paper that is cited once in an important discovery is less crucial than the scientist who writes many papers which are cited many times in many important discoveries.

While it might be maintained that all of the work referred to in a paper was necessary for the production of that discovery, it does not therefore follow that all the particular men who are cited were necessary for the discovery. Although all scientists are replaceable in the sense that other scientists would eventually duplicate their discoveries, some scientists have many more functional equivalents than others. For example, it would be relatively difficult to replace the work of a Murray Gell-Mann, but not as difficult to replace the work of a scientist who is cited once in one of Gell-Mann's papers. If the less distinguished scientists have many functional equivalents, then so do the many laboratory technicians and staff workers who often perform vital tasks in the making of scientific discoveries. We are not saying that the tasks are unnecessary—only that there are many people who could perform them.

To use a comparison which may clarify the point, garbage men perform socially useful and necessary functions; without them a complex industrial society would not function very smoothly. A prolonged strike of garbage men would probably create more chaos in the short run than a strike of teachers, social workers, or even perhaps nurses and doctors. Yet the job that garbage men do could be performed by the National Guard, whereas the jobs performed by professionals could not be done by untrained people. The reason that the garbage man is given little prestige in the hierarchy of occupations is not only because of the lower salary and poorer working conditions that he has relative to a doctor, lawyer, or a scientist, but also because he has many more functional equivalents in the

19. Ibid.

social system than doctors, lawyers, or scientists do.[20] It is far easier to find replacements for the individual garbage man than for the individual scientist or doctor. The same principle operates within a single occupation. Within science some men are more easily replaced than others. We suggest that it may not be necessary to have 80 percent of the scientific community occupied in producing 15 or 20 percent of the work used in scientific discoveries of significance, when perhaps only half their number could produce the same work.

2. A second possible criticism of our analysis is that we have dealt with only one generation of influence.[21] Untested here is the possible "filtration" of ideas from the lower to higher levels of the stratification system. This filtering process may take a number of "generations" of papers before the low-impact papers have an influence on important discoveries. Further, in the process of filtration a minor contribution may be entirely absorbed by the paper from the next generation that makes use of it. Thus only a single citation might be necessary for a piece of work to become part of the stockpile of knowledge. A minor contribution, then, might ultimately have an effect on the production of a great idea through a "great chain" of papers. The first links in this chain would be concealed from our vision because they were not cited by papers of a later generation. What is clearly called for is a study of the sociometrics of multiple generations of papers, in which we examine the number of scientists who are added to the list of those who influence discoveries as new generations of papers are added.

We are currently conducting a study in which we are tracing back patterns of influence. As we go back, we add new names to the matrix. But in line with the assumption of the "replaceability" of scientists, we would argue that the crucial question is not how many new scientists are added to the matrix, but how many central names are added to the matrix. We might define as "central" a scientist whose name appears three times or more. We would guess that as new scientists are added to the matrix the proportion of central scientists will drop off sharply and soon hit zero. We hypothesize that we will not have to look at many generations of influence before we find that all new names added to the matrix are appearing

20. Davis and Moore, "Principles of Stratification."

21. Another criticism that has been made of our thesis is that there may be turnover in the productive scientists within one cohort, that is, scientists who are productive and highly cited at T_1 may not be highly cited at T_2. This criticism is not supported by the data; see the discussion of table 1 in chapter 2.

only once. Kessler found the same pattern in his study of citations in the *Physical Review:* 95 percent of the references were to articles published in the *Review* itself plus fifty-five other journals. He suggests:

> The same list of 55 journals . . . will account for the majority of references year after year. The remaining 5 percent of the references is to a large and ever-growing list of rarely used sources. . . . This list has no stability in time; each new volume examined is destined to carry 96 percent of the references in the subsequent 35 volumes. As we examine those subsequent volumes, 78–96, it is clear that although the list of new titles never ends, their contribution to the total reference literature is comparatively small.[22]

3. A third possible criticism of this chapter could be that we have considered only the research function of scientists. As we pointed out above, scientists can make important contributions to the advance of science through excellent performance in other roles, such as teaching and administration. However, just as it would be incorrect to ignore these important roles, it would also be an error to assume that there is a necessary separation of them from the research function. It is possible that the same scientists who produce the most significant research are also doing the most significant teaching and administration.

Let us look at the teaching function performed by scientists. If the assumption is correct that it is primarily elite scientists who contribute to scientific progress through their research, then we should be primarily concerned with the teachers of future members of the elite. We know from qualitative sources and statistical studies of Nobel laureates, National Academy members, and other eminent scientists, that the great majority of scientists who reach the elite strata are trained by other members of the elite.[23] In fact, 69 percent of current members of the Academy and 80 percent of American Nobelists received their doctorates from only nine universities. It might be facetiously asserted that the best way to win a Nobel Prize is to study with a past laureate. Analysis of the graduate schools attended by physicists whose work is heavily cited indicates that a large majority of scientists who turn out to be productive get their

22. M. M. Kessler, "Some Statistical Properties of Citations in the Literature of Physics" (Massachusetts Institute of Technology, 1962).

23. Zuckerman, "Nobel Laureates in Science: Patterns of Productivity, Collaboration, and Authorship."

doctorates at the top twenty graduate departments. We would not claim that unproductive scientists teaching at low-prestige institutions serve no function—for one, they may serve the truly important function of educating nonscientists in the objectives and methods of science. There is little evidence, however, that they contribute to the progress of scientific research through their teaching.

4. Another possible criticism could be that even if all our hypotheses were supported by the necessary extensive future inquiries, we would still be left with a critical and difficult problem before any policy implications that may be implicit in these data could possibly be acted upon. We would still have to identify correctly those scientists who will go on to produce important scientific discoveries. We would need a set of accurate predictive measures that could identify at an early age those students with creative potential who would produce truly significant discoveries. Although beyond our current capabilities, we believe that this would not be as difficult a problem to solve as it at first seems. As pointed out above, the majority of scientists who contribute to scientific progress are educated at a small number of graduate institutions. Probably most of the exceptions go to institutions of lower prestige for personal reasons rather than because they were rejected by the leading departments. If there were only about twenty graduate departments in each field, all people showing any talent and interest in that field would by necessity have to apply to one of these institutions. If twenty departments each admitted between twenty-five and fifty new graduate students each year, thus potentially reducing by a factor of two the number of doctorates granted, it is unlikely that many students would be denied access to graduate education who had the potential to make important scientific contributions. For example, if twenty graduate departments of physics admitted only fifty students a year, forty of whom were to receive their doctorates in due course, these twenty departments would produce eight hundred Ph.D.'s each year, or about half the total number of American physics doctorates awarded in 1970.[24] A reduction in the absolute number of training centers would not imply a reduction in the competition between these universities for talented researchers or students.

It is an all too well-known fact that new Ph.D.'s in science, especially in physics, are having a difficult time finding jobs. Most

24. National Research Council, *Summary Report, 1970 Doctorate Recipients from United States Universities.*

projections of supply and demand for scientists are not optimistic.[25] One way to handle this inequity in supply and demand is to sharply cut back the number of Ph.D.'s being produced. The data we have reported lead to the tentative conclusion that reducing the number of scientists might not slow down the rate of scientific progress. One crucial question remains to be answered. If the number of new Ph.D. candidates is sharply reduced, will there be a reduction in the number of truly outstanding applicants or will the reduction in applicants come from those who we would now consider borderline cases? This is not a question of social selection, as we believe it possible for academic departments to distinguish applicants with high potential. It is a question of self-selection. A reduction in the size of science might motivate some very bright future scientists to turn to other careers. The ability of an occupation to attract high-level recruits depends to a great extent on the prestige of the occupation, working conditions, and perceived opportunities in the occupation. We, of course, do not intend to suggest the advisability of any policy that would either reduce the prestige of science or the resources available to scientists. What we are suggesting is that science would probably not suffer from a reduction in the number of new recruits and an increase in the resources available to the resulting smaller number of scientists. Perhaps the most serious problem that science faces today in recruiting is the perceived reality that there are few jobs available to new Ph.D.'s. Reducing the size of science so that supply is more in balance with demand, may ultimately increase the attractiveness of science as a career.

25. Dael Wolfle and Charles V. Kidd, "The Future Market for Ph.D.'s," *Science* 173 (August 1971):784–93; Allan M. Cartter, "Scientific Manpower for 1970–1985," *Science* 172 (1971):132–40.

9 UNIVERSALISM AND ITS CONSEQUENCES IN SCIENCE

If the social system of science operated on completely rational and universalistic principles, quality of role-performance would be the sole criterion upon which scientists would be evaluated, and all scientists producing high-quality work would be rewarded regardless of their other characteristics. Their nonscientific statuses, such as age, race, sex, and religion; their scientific origins (where they earned their doctorate); and their location in the scientific social structure would all have no influence on the amount of recognition received. Such a system would be a utopia. It is obvious a priori that science is not a utopian institution. The main purpose of the research reported in this book has been to investigate the extent and ways in which science departs from its rational and universalistic ideal. All our studies have focused on this one problem area. The general conclusion of our research has been that science does to a great extent approximate its ideal of universalism. In almost all cases where science departs from the ideal we find the process of accumulative advantage at work. People who have done well at time 1 have a better chance of doing well at time 2, independently of their objective role-performance; the initially successful are given advantage in subsequent competition for rewards. In this chapter we summarize our results and speculate on their significance for science and scientists. We are particularly interested in taking a closer and more critical look at possible alternative interpretations of our findings on universalism in science.

If we consider the end product of the process of stratification in science, we are struck by the relatively high correlations between measures of role-performance and measures of rank in the stratification system. For example, the number of awards received by a physicist is correlated ($r = .57$) with the quality of his work as measured by citations. Those scientists who occupy the most prestigious instrumental and symbolic positions stand out as a group whose contributions to the advance of knowledge has been considerably above average. There can be little doubt that the most eminent scientists are those who have done the highest-quality scientific work. However, from these high correlations between role-performance and rank we cannot infer that science is completely universalistic. To make this point let us draw an analogy with the general society.

Stratification researchers have found a high correlation between the mean I.Q. of members of an occupation and the occupation's prestige.[1] Members of high-prestige occupations have on the average considerably higher I.Q.'s than members of low prestige occupations. From this empirical finding two quite different and contradictory conclusions can be drawn, depending upon the assumption adopted. If it is assumed that I.Q. is an adequate measure of native ability, then we would conclude that the United States is an extremely universalistic society. The distribution of rewards would be dependent upon ability, the true hallmark of a meritocracy.[2] However, what if we were to assume that I.Q. is not a measure of native ability at all, but merely an artifact of initial location in the stratification system? Then we would conclude that the relationship between I.Q. and the distribution of rewards is a spurious rather than causal correlation. It would not be high native ability but social advantage which would lead to the occupation of a prestigious position. The system would be far from universalistic. The same interpretative problem is faced in analyzing the meaning of the correlation between role-performance and recognition in science. To what extent is the correlation a result of a self-fulfilling prophecy—does the social system of science set up the conditions for the success of some members and the failure of others?

An adequate answer to this question would probably require a real-life experiment which would be difficult to perform. We would have to take a random sample of science students, give them all the resources and advantages possible, and compare their performance with members of a control group who followed normal scientific careers. Without carrying out such an experiment it is quite difficult to decompose the influence of initial ability and environment on recognition received. Since such an experiment is at least currently impractical, we must base our tentative conclusions upon studies

1. O. D. Duncan, D. L. Featherman, and B. Duncan, *Socioeconomic Background and Achievement* (New York: Seminar Press, 1972); John B. Miner, *Intelligence in the United States* (New York: Springer Publishing Company, 1957).

2. Young, *Rise of the Meritocracy.*

using statistical controls. Before reviewing our findings, let us examine the ways in which self-fulfilling prophecies may be set up in science.

ACCUMULATIVE ADVANTAGE AND SELF-FULFILLING PROPHECIES

The high level of universalism in science only indicates that work will be judged upon its merits. It does not insure that all scientists will ultimately have an equal chance of producing high-quality work and therefore of receiving recognition. Due to initial location in the social structure, labelling processes, and accumulative advantage, some scientists have a greater probability of succeeding than others. Once those in positions of power and authority have defined which younger scientists have "talent," the probabilities of success may shift markedly in favor of the chosen. Indeed this labelling process has all the features of a self-fulfilling prophecy.[3]

How might labelling work to produce a self-fulfilling prophecy? In graduate school, certain students are labelled as being "bright" and "promising." They usually become the students of the most powerful and eminent professors. As graduate students, they are given access to greater resources and often have the opportunity to publish papers with their mentors. Perhaps even more important, they pick up self-confidence and the belief that they have what it takes. The "knighted" students of the most eminent professors are also most likely to receive first-job appointments to prestigious academic departments or research laboratories.[4] At these research centers they again have resource advantages and find it easier to publish.

At this point another step in the process of accumulative advantage may take place. A young scientist at a leading research center publishes a paper which is labelled "interesting." He then has even greater resources made available to him as he becomes more integrated into the establishment. It is now true indeed that he has great advantages over his colleagues at graduate school who were not labelled as "bright," were not fortunate enough to be sponsored by an eminent professor, and who did not pick up the self-confidence needed to do high-quality research. The probabilities of the favored young scientists receiving support for their research,

3. For a discussion of various aspects of labelling theory, see Howard Becker, *Outsiders* (New York: Free Press, 1963); Edwin M. Schur, *Labeling Deviant Behavior* (New York: Harper & Row, 1971).

4. This section has benefited from discussions with Eugene Weinstein and Jerome Singer.

working in social and intellectual environments that are conducive to new discoveries, and collaborating with other imaginative scientists, are now no longer the same as those for their less-favored colleagues. The chosen are given an advantage over the others. The personal history of most Nobelists testifies to this process.[5] A major portion of them studied with previous winners of the prize. In fact when we identify the individuals who nominate scientists for prizes, awards, grants, and positions, it often turns out to be the recipients' teachers. In one sense then, it is no surprise that laureates have been the students of other laureates, since former winners are among the few scientists who are asked to make nominations for future prizes. Zuckerman summarizes these points:

> Young scientists who had the opportunity to learn "what matters" from eminent investigators start their careers with an early advantage. They continue to benefit from this experience not only because it has improved their research skills but also because it gives them visibility to those who make decisions on jobs, fellowships, research money, and awards. These young men have, so to speak, been properly launched and however universalistic in allocating resources and rewards the "gatekeepers" wish to be, they are bound to favor candidates they know and respect over others who might be equally able but who are unknown to them.[6]

Does the fact that labelling occurs, that some scientists do benefit from accumulative advantage, and that initial location in the social structure does affect a scientist's life-chances negate the conclusion that science is basically an equitable institution? In answering this question we must first consider the correlates of being labelled as "bright" or "pedestrian," or of being admitted to a high-prestige graduate department. This is a question that some adherents of labelling theory often conveniently ignore. In order to conclude that science is inequitable, we would have to find that there are no differences in the scientific abilities of those labelled "bright" and those labelled "pedestrian"—or worse still, that the favored students have less scientific ability than the unfavored. To answer this question empirically is difficult because it is almost impossible to get an independent measure of untrained scientific ability. We do know that graduate fellowships are awarded primarily on the basis of grades received as an undergraduate and scores on the Graduate

5. Zuckerman, *Scientific Elite*.
6. Zuckerman, "Stratification in Science."

Record Examination. We also know that the brighter students, at least as "brightness" is measured by I.Q. tests, are more likely to attend and receive Ph.D.'s from the top departments. In physics and mathematics the mean I.Q. of men receiving Ph.D.'s from "distinguished" departments is one standard deviation above the mean I.Q. of men receiving degrees from the lowest-prestige departments.[7] Furthermore, we have the testimony of many scientist-teachers that it is relatively easy to pick out the best students—they generally stand out like "sore thumbs."[8] The problem in most graduate departments is a dearth rather than a superabundance of talented students. Because really good students are rare, the probability of a mediocre student being labelled "bright" may be considerably higher than that of a bright student being labelled "pedestrian."

It still might be argued that bright students who are rebellious or who have difficult personalities might be penalized. Without doubt there are some individuals in all institutions who do not achieve their potential because they do not fit in with the institution, or because the institution does not fit in with them. In science this is likely to be an unusual occurrence. Many of the most famous scientists have private and sometimes public reputations as being "difficult." In fact, it is sometimes maintained that there is an inverse correlation between scientific ability and being "a nice guy." Furthermore, scientists tend to place greater importance on instrumental than on expressive behavior. Most scientists would probably rather have a bright rebellious student than a dull docile one. Finally, the teachers who are least likely to feel threatened by rebellious students are probably the most eminent: teachers who are secure in their own abilities are more likely to tolerate challenging students.

The extent to which ego strength and self-confidence is enhanced by being labelled "bright" in graduate school is also problematic.

7. Jonathan R. Cole, "American Men and Women of Science" (Paper presented at the annual convention of the American Sociological Association, Denver, Colorado, August, 1971).

8. Anyone who doubts this should perform a simple experiment. Ask a colleague at a graduate school to send you a set of final examinations or term papers. Read them through and try to identify the papers displaying the most promise. Then check to see how the students are locally rated. Between fields, of course, there may be substantial differences in the predictability of which students will produce outstanding work and which will not. It seems likely that at the graduate school level it is far easier to predict the intellectual capabilities and potential of students in physics than it is in sociology. One reason for this is that by age thirty a physicist has already "matured"; it may take far longer for the maturation of a sociologist.

Answers to two questions are needed. First, to what extent is self-confidence established prior to entrance to graduate school? In fact, might not self-confidence help to determine which label is put on a student, rather than be determined by such a label? Second, in what proportion of cases does being labelled as bright actually serve to increase self-confidence? It may be overlooked that sometimes being the student of a demanding eminent professor can put a strain on one's self-confidence. Some eminent scientists are known for being very hard on the students they have picked out and labelled as bright. In some proportion of cases, working with such a man may weaken rather than enhance self-confidence.

What about students who are bright but have intellectual views which diverge radically from those of their mentors? If Young Turks or brilliant young men with ideas which are imaginative but contrary to accepted belief are initially labelled as "crackpots" and "wrong-headed," then science may cool out a few potentially brilliant men by refusing to support their work.[9] We believe that this is a rarer occurrence than critics of science maintain. First, in natural sciences like physics there seems to be substantial consensus on what is an interesting problem and on the overall paradigm of the science.[10] It would be interesting to give a questionnaire to successful and less successful graduate students in physics, to see if their basic views of the field differ. We doubt that they would. In the social sciences this problem is perhaps greater. A bright sociology student interested in symbolic interaction who mistakenly ends up at Columbia or Harvard, where there is little instruction in this orientation, is unlikely to be happy. But most students have at least some vague idea of what the approach is at different graduate schools. Furthermore, it is not impossible, and sometimes actually happens, that students transfer from Columbia to Minnesota, Northwestern, or Michigan. This pattern occurs frequently in the physical and biological sciences, where students are known to follow their teachers from one university to another. In general the most eminent profes-

9. The inherent problem in such cases is to empirically identify them. They are analytically akin to the discovery of great value which has forever gone unrecognized. It is virtually impossible to estimate how many such discoveries there have been of this type. Just as it is equally impossible to identify brilliant young scientists who have left science because they went against the grain and were denied intellectual, social, and, not least of all, financial support.

10. Another reason why it is rare for students to have radically divergent intellectual perspectives from those of their teachers is that the intellectual views of students are generally molded by their teachers.

sors are probably more likely to be impressed by students who have independent minds than by those who simply mimic their teachers. We must emphasize, however, that these are speculations and are not presented as conclusions.

One other point must be made about labelling and the process of accumulative advantage. Receiving positive labels will not assure a student a position of eminence unless he in fact does go on to produce high-quality work. Because labelling and accumulative advantage gives one greater opportunity to produce good work, it is sometimes easy to forget that the work must actually be produced and that many students labelled as bright fail to produce, and never attain eminence. Even students who are highly talented and so-labelled must perform. What happens if a not-so-talented student is mistakenly labelled as bright? It is unlikely but nonetheless possible for a relatively mediocre student to be labelled bright and, under the sponsorship of an eminent mentor, to get a good first job. But what will happen if this man does not live up to his billing? He is likely to find himself downwardly mobile. It is difficult to hide incompetence in the "competitive world of the pure scientist." To become eminent in science, one's work must be valued by a wide range of scientists. No matter how "good" a student is, he must prove himself to colleagues all over the country—to scientists who have no vested interest in labelling him as "bright" and who will usually not hesitate to point out his errors and give him the labels of "trivial" or "shallow." Furthermore, in science it is difficult even for already eminent scientists to rest on their laurels. Instead of remembering good work done in the past, evaluators are likely to ask, "What has he done lately?"

Thus, the labelling process and accumulative advantage are highly unlikely to result in an unqualified scientist becoming eminent. These processes, however, are likely to give greater benefit to the fortunate people who start off with talent and are able to use their talent effectively. The labelling process is probably more likely to have a negative effect on the careers of mediocre students. If we consider two mediocre students—one labelled as "bright" and the other as "pedestrian"—the former is likely to develop greater self-confidence and more likely to sustain the motivation that is required to become even a run-of-the-mill research scientist. The mediocre student who is not encouraged is more likely to drop out of science or, if he remains in science, to turn away from research.

Thus far we have been examining the ways in which labelling and processes of accumulative advantage might act to set up a self-fulfilling prophecy in science. Several of our studies offer data which bear upon this problem. First let us examine the influence of nonscientific statuses on the distribution of rewards. If women, blacks, and Jews were to fare poorly in science, it might be a result of accumulative disadvantage and labelling. However, the data indicate that sex, race, and religion have little if any independent effect on rank in the stratification system after the receipt of the Ph.D. Men and women scientists are, on average, about equally rewarded; there are not enough black Ph.D.'s to even consider the question of racial discrimination after receipt of the Ph.D.; and if anti-Semitism exists in science it has no measurable effect on the distribution of rewards to university scientists. If an equitable situation does exist, it is unnecessary to consider the possibility that a self-fulfilling prophecy may have created such a situation.[11] Although women scientists do not receive significantly less recognition than their male colleagues, they are less likely to enter science and less likely to be productive after receipt of the Ph.D. Explanation of these facts might best be sought in what happens to men and women prior to receipt of the Ph.D. and probably prior to entrance to graduate school, since evidence on graduate admissions and the distribution of graduate fellowships give no indication of discrimination against women.[12]

Although nonscientific statuses appear to have little influence on status attainment in science, location within the social structure of science does influence status attainment. Graduates of top-ranking doctoral departments are slightly more likely than graduates of less prestigious institutions to get better first academic jobs.[13] Once a scientist is located at a prestigious institution he has a better chance of receiving a continuing appointment (tenure), than an equally qualified scientist would have of being upwardly mobile into that department. Although social origin of a scientist has an independent effect on positional success, it has very little if any influence on reputational success.[14] Where accumulative advantage plays its

11. Intrinsic to the concept of self-fulfilling prophecy is the fact that the initially down-graded group actually does worse. If this group does not do worse, the concept of self-fulfilling prophecy is not useful. It should be clear that even though a self-fulfilling prophecy is not at work, discrimination could be. This would occur when an initially downgraded group actually has superior talent but in the end does not do any better than other groups.

12. For a full discussion of this point see chapter 5.

13. See chapter 3.

14. See chapter 4.

greatest role is among those scientists who already have received some form of recognition. The act of recognition increases the perceived quality of the scientist's work, which will in turn bring him further recognition.[15]

Our data lead us to conclude that the single most important variable in influencing the distribution of rewards is the quality of one's work as it is perceived by colleagues. The extent to which the work is deemed useful (that is, highly cited) is the key determinant of perceived quality. It is relatively easy for scientists to tell the difference between work of sharply divergent quality. However, their opinions of work of roughly equal quality may be influenced by factors other than the substantive content of the work itself. Scientists who have graduated from prestigious departments, who present themselves in a favorable light, who are currently employed at a prestigious department, and who have received previous recognition are more likely than their less fortunate colleagues to have their work well-thought-of. We would conclude that producing work of *extremely* high quality is both a necessary and sufficient condition for scientific success. Producing work of high quality is a necessary but not completely sufficient condition for success. Location in the social structure, through the process of accumulative advantage, helps the producers of high-quality work.

It is relatively easy to demonstrate that scientists who have received recognition are competent and deserving. But what of the competent and deserving scientist who goes unrecognized because he is incorrectly labelled or because of his initial location in the social structure? Although we fully recognize that it is impossible to answer this question conclusively, the data that we do have lead us to conclude tentatively that this is probably not a serious problem in contemporary science because it appears to be a relatively rare occurrence. The system seems to take many precautions to protect itself against this type of error in evaluation. Consider first the acceptance rates of the journals in the natural sciences. As we have several times pointed out, virtually all papers that are submitted are published. The outcome of this high rate of acceptance is that the great bulk of published papers are infrequently cited. It might be argued that, although most papers can be published, only a small, favored group of scientists have access to the resources and facilities necessary to produce papers. This might be a serious problem in fields like physics where expensive equipment is often required

15. Ibid.

to do research. Although governmental research facilities such as those at Brookhaven National Laboratory, Stanford Linear Accelerator, and Argonne National Laboratory are open to any scientist whose proposal is approved by a panel of peers, it is, of course, possible that the judgments of the peer review panels may in part be influenced by factors other than the substantive content of the proposals. Further research is needed on the criteria employed by gatekeepers in the distribution of resources.

To protect against the error of overlooking qualified talent, granting organizations such as the National Science Foundation have made concerted efforts to seek out proposals from scientists located at nonprestigious departments and from young scientists who have not yet made their mark.[16] Some granting agencies have gone so far as to set up systematic procedures to search for unrecognized talent as potential recipients of fellowships.[17] Finally, consider the fact that social origins of scientists, or where they earn their Ph.D., explains only a small portion of the variance on ultimate rank in the stratification system. We conclude that it is unlikely that there is a substantial number of scientists who possess talent and are completely overlooked. It is far more likely that, because of location in the social structure, some talented scientists gain slightly more recognition than other talented scientists. If we consider two scientists who have high and roughly equal ability, the one graduating from a top department will end up at a "distinguished" department and the one graduating from a less prestigious place will end up at a "strong" department. They both have done well, but one slightly better than the other.

What effect does the process of accumulative advantage have on individual scientists and science as an institution? The answer to this question is strongly dependent on whether there is a positive correlation between initial ability and placement in positions where the individual can benefit from the process of accumulative advantage. Let us make the unproven assumption that there is such a correlation. Is it then unfair to give greater resources to those with the greatest creative potential? Although this might be in some sense "unfair" to individuals, it may be the only rational way to operate for institutions with limited resources. If professors were

16. Information has been conveyed to us in private communications with various scientist "gatekeepers."

17. One organization that has done this is the Center for Advanced Study in the Behavioral Sciences.

to divide their time equally between bright and pedestrian students, if granting agencies were to indiscriminately make grants, disregarding the past performance of the applicants, the rate of scientific progress would probably be retarded. Only if resources are unlimited can an institution such as science afford to treat everyone equally.

In conclusion, the significant point is that a self-fulfilling prophecy is not necessarily detrimental for a social system. If the prophecy is based upon a correct initial assumption of superior creative potential, then all turns out well. Negative consequences only obtain if there was an incorrect prediction of the talent of the young scientists. What actually happens is clear, however. Science does define some students as "promising" and others as "run-of-the-mill" and then systematically sets up the very conditions that produce the expected and predicted consequences. The question is, of course, whether those who do not have the advantages of facilities and resources would do anything with them if they had them. Quality is rewarded, but is "quality" more a result of innate ability than a process of accumulative advantage? Of course, if both variables are strong independent contributors to productivity, as we believe they are, then science is operating at its most efficient level. For this would indicate that the processes of self- and social selection are operating to identify scientific talent, assign it to positions where it can best be used, and place at its disposal the resources that are needed to produce new discoveries. Research is clearly needed on the actual operation of accumulative advantage. Attempts at isolating the independent effects of quality and sponsorship on career patterns should turn up results of great interest.

This discussion has concentrated on possible advantages accruing to individuals who are labelled as "bright" early in their scientific careers. We have been concerned with the relationship between self-fulfilling prophecies and accumulative advantage which can result in the "chosen" prospering in a domain of scientific riches while the neglected go poor in less fertile regions of the community. But there is another, nontrivial, aspect to the labelling process. There are some circumstances under which being labelled favorably will have negative consequences on the young scientist's career. As Merton has noted, under some conditions "self-fulfilling prophecies" can be transformed into "suicidal prophecies."[18] In

18. Merton, *Social Theory,* p. 423n.

short, the prediction itself of future eminence can alter the conditions and the subjective definition of a social situation and produce an outcome which was both unanticipated and at variance from the predicted outcome.

Social life is full of examples of false prophecies due in part to the prediction itself. In athletics, for instance, there are numerous clear examples in which an individual's early performance is followed by encomiums which in turn produce changed conditions for the athlete and consequent decline in levels of performance. Such self-negating predictions can arise in science when public or subterranean pronouncements of the imminent arrival of a new star increases psychological pressures on the individual to the point where the heralded scientist perceives his situation with despair, for it no longer seems possible for him to live up to his advance billing. Reacting to these extraordinary expectations, the new "star" suddenly develops research and publication "cramp," reducing his output and thus "proving" the soothsayers fallible. Beyond the psychic consequences of being labelled a potential star, there also can be purely structural conditions which induce "suicidal prophecies." For example, it is not uncommon for young stars to be invited by scientific societies to become members of prestigious committees and to join the editorial boards of major journals where they are given a large number of tasks to perform.[19] These new activities can consume large portions of their time and energy which might otherwise be used to conduct research. Being labelled a bright researcher obviously can overflow into being labelled a bright everything, leading to a subversion of energy. Research ability is assumed to be highly correlated with competence in the other roles that scientists perform.

In summary, the status sets of these "bright young scientists" become greatly enlarged, their role obligations expand, and they are provided with a convenient rationale for scaling down their research effort and not testing themselves against the expectations of the scientific community. Of course, most brilliant young scientists either resist the temptations of these prestigious appointments or simply make the time for additional obligations without reducing their research effort. In such cases the original prophecy does not

19. Within the last few years, there has been a sharp increase in the efforts of boards of directors, scientific journal editorial boards, and prestigious committees to broaden the base of representation on their panels by coopting able scientists among the young, the minority groups, and women. Able young scientists face the lure of attractive titles perhaps more today than in the past.

fail. In some cases, however, having altered the social and psychological conditions which initially attended the designation "star," the research performance of the individual scientist probably does fail to live up to expectations. There continues to be a problem here of determining whether the failure to live up to expectations is due to the influence of the prophecy or to an incorrect definition of the initial talent of the individual. Far more research is required on the frequency of "suicidal prophecies" and the conditions under which such predictions are likely to become self-defeating.

The major thrust of this book, then, is that science, more than most social institutions, approximates the ideal of universalism in its assignment of individual scientists to positions in a highly stratified social structure. Within the economy, the political system, and even within other professions, there seems to be a greater discrepancy between the ideological commitment to universalistic standards and the actual operation of the system. While there are a number of identifiable abridgments of the principle of universalism in science, deviations from the norm are generally the product of a calculus in which individuals are unconsciously "sacrificed" for the purpose of advancing the entire scientific community towards its research goals. We now turn our attention to an examination of a few of the social consequences for science of the structure that has been identified.

FUNCTIONAL AND CONFLICT THEORIES OF STRATIFICATION

We have considered two divergent theories of stratification, the functional and the conflict theories. Although all our studies together do not enable us to determine conclusively which of these two theories is more applicable to scientific stratification, as no theory can in fact be "proven," they offer us the opportunity to make some tentative speculations. The functional theory posits that those positions which fulfill the most important (the most highly valued) functions will be the most heavily rewarded.[20] Our studies indicate that this is true in science. There can be little doubt that, in science,

20. There have been numerous critiques of the functionalists, particularly the argument of Davis and Moore that the relative functional importance of occupational statuses and roles within society determine the rewards accruing to practitioners. The critiques have focused primarily on the difficulty in measuring functional importance. We shall not enter this debate, except to note that within a single set of occupations, like scientific occupations, it is probably far easier to establish the relative functional importance of certain roles, than it is for the society as a whole. For references to critiques of Davis-Moore, see chapter 3, n. 41.

contribution to the advance of knowledge is the most highly valued activity, and our data show that those scientists publishing the best work are the most heavily rewarded.[21] The functional theory also postulates that rewards will vary inversely with the number of qualified people capable of carrying out a task. Our studies indicate that this is also true in science. Scientific progress depends heavily upon the work of a very small elite.[22] It is probable that few people have the ability to enter this elite; there are many more people who have the ability to carry out the technical tasks upon which some of the elite depend in getting their work done.[23]

Conflict theorists would probably not deny the facts that our research has uncovered, but they would question the interpretation we have given the facts. They would not argue that rewards are not given primarily to the men producing the most cited work; but they would question the processes through which some work is heavily cited and other work is not. Let us consider systematically several points in a possible conflict critique of the functional theory of scientific stratification. Since conflict theorists would emphasize the importance of accumulative advantage, at points we must necessarily repeat some of the above discussion.

Critique 1: Resources enabling a scientist to do the research leading to publication are controlled by a power elite who distribute the resources to their friends, students, and others who will serve their interests, ignoring equally talented scientists who are not in the inner circle.

The only way that this aspect of the conflict theorists' critique could be tested is to devise a measure of native scientific ability which does not depend upon performance. The only such measures that we have are scores on I.Q.-type tests. These tests are unsuitable predictors of scientific ability for several reasons. For one, scientists have such high I.Q.'s that the scale is probably a fairly unstable

21. See chapter 4.

22. See chapter 8.

23. One criticism that has been made of this line of reasoning is that if all the current elites in a field would suddenly retire, a new set of elites would emerge to take their place. This argument is used to demonstrate that current nonelites have the ability to be elites. The argument is correct in assuming that every science needs an elite but, to us, seems to be incorrect in assuming that people not currently in the elite could be transformed into elites without doing the science some serious harm. If a science cannot recruit enough people with the capability of occupying positions of leadership, the science will most likely decay, or stagnate. It is, of course, true that our data, which show that very few scientists contribute to progress, do not prove that there are not more people capable of making contributions.

measure of native ability for this select population. Secondly, I.Q. tests do not measure specific types of ability. If we did use I.Q. tests as a measure of native ability, we would probably conclude that resources are for the most part given to the more deserving. As we pointed out above, the mean I.Q.'s of students at top-ranking departments are higher than those at low-prestige institutions. It appears that the density of bright students is heavier at the top departments than at others.

Given that it is difficult, if not impossible, to devise a reliable measure of native scientific ability, we must attempt to resolve this question in other, less precise ways. One way to do this would be to take fields of science that do not require expensive equipment to conduct research, and see if productivity distributions are the same in these fields as they are in those fields which do require expensive equipment. If in fields not requiring expensive equipment we were to find that the bulk of good research is done by a small minority of scientists who cluster predominantly at the major institutions, this would provide some evidence that a high rate of publication is not primarily a function of the availability of resources. Recent research on the reward system of physics, chemistry, biochemistry, psychology, and sociology has shown remarkably similar results. Fields like sociology and psychology, which certainly require less expensive equipment than physics, show productivity distributions almost identical to those in the hard sciences.[24] Although it is true that the uneven distribution of resources gives an "unfair" advantage to some scientists over others, we would hypothesize that a sharp reduction in the support of research or a change in the mode of distributing resources might reduce the total number of papers published but would not change the shape of their distribution or the identity of their producers.

We are not asserting that no or only a few talented individuals fail in science; we believe that many talented people fail in it. But the reasons for their failure cannot generally be traced to inequities in the social structure of science. Failure among talented scientists is probably more often a result of lack of motivation or of other psychological factors which hinder productivity. To test this hypothesis we need to do case studies of students who were considered to be very talented in graduate school but who later either dropped out of science or failed to succeed.

24. Stephen Cole, "Scientific Reward Systems."

Critique 2: The journals are controlled by a small clique of powerful scientists. Scientists who are not members of this clique find it more difficult to publish.

Available data on acceptance rates of journals indicate that this is empirically unlikely.[25] Almost any paper can get published in physics. The great majority of scientists publishing in *Physical Review* can in no sense be seen as eminent or part of an "inner circle" in science.

Critique 3: Once papers are published, the power elite controlling physics will ignore those published by outsiders and cite those published by insiders. Heavy citation results as much from membership in a social group as from the substantive content of the work.

Several of our studies offer data which suggest that this criticism is probably untrue. First, consider the studies of the Matthew Effect. These studies showed that if a scientist produced a good paper it would be quickly recognized and utilized regardless of the scientist's location in the stratification system.[26] Scientists who have received the most prestigious awards, entitling them to membership in the elite, are no more likely to have a good paper immediately utilized than a scientist who has received no award and is definitely not a member of the elite. It may be argued that this research depends only upon papers that were at some time recognized as high-quality, and that many more papers by nonelites are likely to go totally unrecognized. We doubt this argument is true and invite a test of its validity. An appropriate test might be to take a *random* sample of highly cited papers in a field and a *random* sample of papers receiving few or no citations in the same field and then have the papers evaluated by a panel of experts with the names of the authors removed. The difficulty, of course, would be that many of the highly cited papers would probably be known to the experts. Evaluations should only be taken of papers the expert is unfamiliar with but competent to judge.

A second set of data, those on utilization of research (presented in both chapters 6 and 8), suggest that citations are not primarily a result of location in the social structure. It happens that members of the elite are not much less likely to cite nonelites than nonelites themselves. Also, nonelites cite their more eminent colleagues almost at the same rate as they are cited by elites.[27] If citation prac-

25. Zuckerman and Merton, "Patterns of Evaluation."
26. See chapter 7.
27. See chapter 8.

tices are determined by social location rather than substantive content, why do not nonelites cite the work of their status peers more often? Conflict theorists might argue that the scientific lower classes have "false consciousness," that they have internalized the world view of the scientific "ruling class." This brings us to the last point we shall consider in a possible conflict-theorists' critique of the functional theory of scientific stratification.

Critique 4: There is not more social conflict in science because the means of communication and the socializing institutions (graduate schools) are controlled by the elite; most scientists develop false beliefs that the reward system is equitable and that their interests are not opposed to those of the elite.

The question that needs further empirical research is whether scientists at all levels of the stratification system do perceive the reward system as equitable and universalistic, and how they view and explain their own position in the stratification system.[28] We would hypothesize that most scientists would see the system as fundamentally equitable and that low-ranking scientists will be no less likely to see it in this way than their high-ranking colleagues. It is possible that low-ranking scientists may do more griping but without any deep-seated sense that they have been treated unfairly by the system. If this turned out to be the case, it would be incumbent upon the conflict theorists to demonstrate the mechanisms through which inequity is viewed as equity. When a group of people hold views which you prefer they did not, it is easy to assert that they suffer from "false consciousness." But to make false consciousness a scientific rather than an ideological concept we must specify in detail the way in which views of the world become distorted.

Clearly, we do not see the conflict orientation as offering substantial insight into the social organization of modern American academic science. The data that we have do not offer much support for the conflict perspective, although it is difficult to demonstrate this conclusively. The data seem to offer considerably more support for the functional orientation. One part of the functional theory which deserves further development is the functional necessity for leaders, or stars, in science. Davis and Moore attempt to explain inequality in the distribution of rewards by the necessity to motivate

28. We do not have data on the attitudes and opinions of scientists variously located in the social system. Therefore, we do not have precise indicators of how they feel about the application of universalistic standards or if they think that universalism is practiced in modern science.

individuals to fill positions involving the most valued functions and the most difficult performance.[29] They concentrate on the necessity of getting the most talented people to fill positions rather than on the necessity of having differentially ranked positions. If all people were inherently motivated to do their best, the Davis and Moore theory would not hold. We believe that at least in science the unequal distribution of rewards serves other necessary functions as well as motivation. If all scientists had the "sacred spark" (an inner drive to do science for its own sake), the system would still need stars—leaders who have received an unequal share of recognition.

We are hypothesizing that in order for a modern science to function effectively it must meet certain functional requirements. Although we are a long way from demonstrating this, we would like to speculate on the functional necessity for stars, or scientists who have received a great deal of recognition. Stars and the unequal distribution of recognition needed to create stars serve two primary functions. The first function is motivational. Some scientists do much better work than others and by receiving widespread recognition become role models who encourage others to follow suit. The second function of stars is that they provide science with leaders who can exercise legitimated authority. To clarify this point, let us imagine a system in which all scientists had the sacred spark (stars, therefore, would be unnecessary for motivational purposes) and positions of authority were filled by lot. We believe that such a system would break down because the leaders would not be granted legitimacy. Leaders serve two primary functions in science: the distribution of resources and the setting of the consensus. Let us consider the distribution of rewards function. A large number of scientists applying for grants and fellowships are turned down. When a scientist is turned down, he may grumble to himself but he is generally not motivated to make any public complaint. This is probably because we grant legitimacy to the people who serve as gatekeepers. We feel that gatekeepers have earned the right to their position through their role-performance. It is probably easier to accept rejection if the person making the evaluation is granted legitimacy. If leaders were chosen by lot, the rate of social conflict over the distribution of resources might increase.

Now let us consider the consensus-setting functions of leaders. Scientific progress is in part dependent upon the maintenance of intellectual consensus. If scientists do not have a common view of

29. Davis and Moore, "Principles of Stratification."

the problems and methods of their discipline, knowledge will not cumulate.[30] Every effort will be an isolated, individual one rather than a part of the communal enterprise. Who determines what the consensus is? The consensus is probably set by the intellectual leadership of the discipline. Leaders do this through the evaluation process. When a Nobel Prize is awarded, it not only rewards a scientist and thus creates a role model, it tells other scientists what is considered good work. If we eliminated leaders who evaluated work, consensus would gradually break down, as every individual would be faced with the impossible task of making countless evaluations of other scientists without any guidelines. If evaluations of work were made by leaders chosen by lot and thus lacking legitimacy, consensus would probably break down even more rapidly. We are unlikely to accept intellectual evaluations made by people who we feel are unqualified to make them. We cite an example from the recent history of sociology. Prior to the establishment of the review journal *Contemporary Sociology,* the book review editors of the *American Sociological Review* decided that they would choose reviewers from the ranks of the little-known. This practice led to an increase in complaints about the reviews, even though there was probably no higher proportion of negative reviews than there had been in the past. The difference was that the little-known reviewers did not have the legitimacy that is required for evaluations to be accepted. This is not simply a case of the same evaluation being given greater weight if made by a star. Evaluations made by stars and unknowns are not the same in content precisely because stars have been selected for their intellectual capabilities. Leaders who were selected on any other basis would not be effective, because of the absence of both symbolic authority and substantive ability. Because of the dual functional necessity for legitimated leaders, we would speculate that the stratification systems of the various sciences would be structurally organized in roughly the same way, regardless of their state of cognitive development or the degree to which stars have naturally emerged. A field which does not have "natural" stars will have to create them. The created stars will be the best people available in the field at the time even though their "absolute" contributions may not be as great as those of the stars who have naturally emerged in other fields or of the stars in their own field who preceded them.

30. Of course, it is possible to have several coexisting paradigms in a field. This is generally the case in the social sciences.

FAILURE AND DEVIANT BEHAVIOR

One conclusion of this book is that science is dominated by a small, talented elite. All major forms of recognition—awards, prestigious appointments, and visibility—are monopolized by a small proportion of scientists. Furthermore, we have concluded that the research of the majority of scientists contributes little to scientific advance. The question we shall consider here is how universalism and elitism in combination affect the average scientist.

Few institutions are as performance-oriented as science. Although the gatekeepers in most occupations probably would claim that rewards are based on an evaluation of performance, the level of universalism is in fact a continuum. Physics would seem to be at one end of this continuum. Even in the most prestigious American profession, medicine, success is probably less a function of ability than it seems to be in science. In fact there would seem to be little agreement on what constitutes being a "good" doctor. To be a high-prestige physician does not seem to depend simply on intellectual capability and clinical skill, but also on highly intangible factors such as capacity for "understanding the problems of patients" (known in its degenerate form as "bedside manner"). Consequently, the physician who did not do particularly well in medical school can still hope to succeed in the profession by earning a high income, relying on his personal attributes rather than on his rigorous professional knowledge. Of course, these expressive abilities would be less important when we examine rewards to research physicians, who are closer to being pure scientists.

The reward system in the American business community seems to contrast sharply with that found in science. Although systematic data on the operation of the reward system in business are lacking, it seems that businessmen are often rewarded for bureaucratic competence and for loyalty to their company. While such characteristics might also lead to local rewards within large-scale scientific organizations, in the larger system of rewards for scientific research these attributes are rarely recognized. Further, consider the incidence of nepotism in American business in contrast to science. Although family capitalism has been waning, the incidence of individual success that may be attributed to family ownership of a business is not by any means rare. In science, there have often been explicit antinepotism rules on hiring practices at university departments. These rules can actually act against the application of universalistic standards, since occasionally a scientist whose work is of the first rank is denied a position at a department solely because a spouse or other relative holds a position there.[31] Of course, these rules are designed to increase the application of universalistic standards of rewards *within* departments.

The application of universalistic criteria aids individuals in evaluating their own performance. Since the criteria for success are clear and well known, scientists have a guide to judge how well they are doing.[32] Contrast the situation of the scientist with that of individuals working in institutions where standards of evaluation are unclear; for example, the situation of the actor or artist. The unrecognized artist faced with hostile critics can easily debit the system for his lack of recognition. And to the extent that the evaluation process is more subjective and unstandardized, the response is plausible. Certainly, artists can point to a large number of historical examples of artists whose work was not appreciated until long after they were gone. While no social system can completely eliminate ambiguity over the criteria of evaluation, science, as much as any other institution, seems to have minimized this ambiguity.

We have no data on whether scientists accept as legitimate the evaluations made of their work. In the absence of a significant number of complaints and until adequate data are collected, we would hypothesize that the bulk of scientists not only accept the legitimacy of the system but also have a fairly clear idea of where they stand in that system. And where most scientists stand is clear. Most scientists have not "made it"; they are relative failures. Fair criteria of evaluation have been employed, and they have not measured up.[33]

How do individuals in a social system like science handle failure? Failure to achieve recognition cannot easily be seen as the result of widespread discrimination within the social system. It is difficult for the scientist to find a reason for his own "failure" in the structure

31. Recently, many universities have relaxed their nepotism rules in response to protest that these rules were negatively affecting the operation of universalistic standards.

32. In a recent organizational study that did not involve scientists, Katz and Kahn showed that a significant proportion of workers simply did not know what they were supposed to be doing or how well they were doing at their jobs. Many workers, more than one-third of those studied, simply did not know how their role-performance was being evaluated. See Daniel Katz and Robert L. Kahn, *The Social Psychology of Organizations* (New York: John Wiley & Sons, 1966).

33. Here we use the term failure in the sense of failure to become eminent. Scientists who are "failures" in this restricted sense may well be "successes" in their local environments.

of the institution. There must, of course, be some scientists who do, in fact, blame the system for their own lack of success. But if a scientist believes the system is equitable, he can only look to the quality of his role-performance for an explanation of his success or failure.

An interesting question remains as to the extent to which individuals in social institutions with less universalistic criteria of evaluation tend to blame themselves and their lack of personal competence or blame imperfect social structure for their failure. In eighteenth- and nineteenth-century America there was probably a greater tendency for all individuals to blame themselves for lack of success than there is today.[34] But the very fact that there are large social protest movements by blacks, women, and the poor suggest that an increasing number of Americans see the imperfect social structure of American institutions as the source of their own lack of opportunities and ultimate "failure."[35]

When a reward system is defined as inequitable, a rationale is provided for violating the norms of the system. In short, the structure of the reward system can produce personal justification for deviant behavior. In science as in other institutions it is difficult to estimate the amount of deviant behavior that occurs. There are occasionally accusations by one scientist that another has stolen his work or not given proper credit for assistance. Some well-known scientists are rumored to be exploitative of graduate students and colleagues. No doubt a few published articles contain falsified data. But such forms of deviant behavior are probably rare. They are certainly rare compared to visible deviant behavior in other institutions. They are rare because they generally are not effective in attaining success and because most scientists seem to have a genuine commitment to the norms.

In a study of close to two thousand academic scientists in five disciplines, Hagstrom found that few scientists perceived that their colleagues were not trustworthy.[36] Only about 7 percent were afraid to discuss their research with other scientists doing similar work.

34. Sigmund Diamond, *The Reputation of the American Businessman* (Cambridge: Harvard University Press, 1955).

35. However, even today many Americans who have not "made it," tend to blame themselves rather than an imperfect social structure. See Stephen Cole and Robert Lejeune, "Illness and the Legitimation of Failure," *American Sociological Review* 37 (June, 1972).

36. Warren O. Hagstrom, "Competition and Teamwork in Science" (Final Report to the National Science Foundation, July 1967), pp. 126–27.

Approximately 13 percent of the scientists said that they had at one time in their career been accused of failing to recognize prior work in their own papers. A slightly higher proportion believe that their own work has been used by others without proper citation.

It would be highly difficult, if not impossible, for an unknown scientist to become eminent through deviant adaptations. It is not easy to "steal" someone's idea. You must first be in contact with the potential victim. Then you must have access to the resources to develop the idea. Noneminent scientists generally have neither the extended contact with eminent scientists nor the intellectual or material resources to operationalize a stolen idea. Instances of accusation of "theft" generally occur among status-peers or collaborators and are more likely to involve alleged failure to give proper credit for a contribution. It is true that occasionally eminent scientists exploit students. But it is difficult to use the exploitation of students as a means to attain eminence. Generally a scientist has to earn a reputation before he can attract more than a few students. Falsifying data is highly unlikely to bring eminence. If the falsified study is unimportant, it will be just as ignored as an unimportant honest study. If the falsified study is important, it will soon be replicated and the "incorrect" results pointed out to the embarrassment and possible discreditation of its publisher.

Perhaps a more important reason for the low rate of deviance in science is commitment to community norms. Most scientists have a genuine intellectual interest in what they are doing. They believe that unravelling the secrets of nature is a highly worthwhile activity. Making a significant discovery gives the scientist a great deal of self-satisfaction. Engaging in deviant behavior to get recognition would rarely provide the self-satisfaction of making a genuine discovery. "Cheating" debases science into a meaningless game. Most scientists would find recognition in such a game uninteresting. Scientists may want recognition but they want it for doing science—for making a significant discovery. Unlike in the business world, where rewards take the form of money, in science the rewards are generally recognition for specific intellectual work. If studies can be falsified and if credit can be stolen, the legitimacy of the institution would be destroyed and recognition would lose much of its value. This once more underscores the fact that, in science, prestige, success, and recognition cannot be separated from the production of significant research. Scientists are motivated by the need to discover the "truth," and the significance of these discoveries is socially validated through recognition. Neither recognition without the discovery (deviance or misallocation of credit) or the discovery without recognition are likely to satisfy very many scientists.

Science is one institution in which there is probably greater commitment to the norms today than there has been in the past. Merton points out that multiple discoveries less often invoke priority disputes and accusations of plagiarism today than they did in the past.[37] This is probably because as science has become increasingly institutionalized it is easier to establish priority and there is greater commitment to the norms. Many scientific fields have set up procedures which make it relatively easy to establish priority. Journals like *Physical Review Letters,* which publish reports of discoveries very quickly, and *Bulletin of the American Mathematical Society,* which publish announcements of forthcoming articles, make it relatively easy to determine priority. Also a highly efficient communication system makes it unlikely that a scientist will not be aware of the work of other scientists working in the same area. Today science is carried out in settings in which it is easier to exercise social control. Scientists who continually violate the norms will be defined as "crackpots." In short, both the institutional structure and commitment to the norms reduce the amount of deviant behavior and the likelihood that deviant adaptations will be used to cope with failure.

Unlike in the larger American culture in which individuals are taught that what one really needs to achieve "success" is persistence and that immediate failures are only momentary and should be viewed as way-stations to ultimate success, it is difficult for scientists to cope with failure by believing that success is just around the corner. Scientists generally either make important discoveries by the age of thirty-five or forty, or they never make them.[38] Scien-

37. Merton, "Singletons and Multiples," p. 463.

38. There have been only a few studies of the relationship between age and creativity in science. See H. C. Lehman, *Age and Achievement* (Princeton: Princeton University Press, 1953); Wayne Dennis, "The Age Decrement in Outstanding Scientific Contributions"; Harriet A. Zuckerman and Robert K. Merton, "Age, Aging and Age Structure in Science," in Matilda W. Riley, Marylin Johnson, and Ann Foner, eds., *A Theory of Age Stratification,* vol. 3 of *Aging and Society* (New York: Russell Sage, 1972), pp. 292–356. We are currently investigating this relationship in our studies at Columbia. The data that we are presently analyzing suggest that if scientists have not made significant contributions by the time they are thirty-five or forty years old, they are very unlikely ever to do so. However, the data show as well that if the scientist does make important discoveries by the age of thirty-five or so, he is likely to continue making important contributions throughout his

tists know that scientific talent is usually apparent at a relatively early age; if major discoveries do not arrive on time, they feel that they will not arrive at all.

In science, frustration of individual aspirations does not seem to lead to a state of anomia. While it would appear rational for a man who has been taught to prize and pursue goals to reject those goals when his own attempts to reach them have been unsuccessful, there is no evidence of widespread alienation in science stemming from a lack of success. Why the level of deviant behavior and alienation is relatively low in science compared to other institutions is a question worth pursuing in future research. Tentatively, we can look to the ideology or value structure of science for possible clues. Young scientists, after all, are not only taught to strive for the production of important ideas and concomitant rewards, they are also taught to identify with the communal aspects of the scientific enterprise.

The normative structure of science has several elements which seem to serve the latent function of reducing alienation resulting from failure. Take, for example, the norm of disinterestedness. Scientists are told that they should do science to discover the truth for its own sake; they should not be concerned with rewards. Many scientists apparently do just this. They enjoy doing science for its own sake with the not too closely questioned assumption of the norm of communality—that in some small but meaningful way the work of all scientists, even the most obscure, contributes to scientific advance. Unlike the general culture, in which failure is interpreted as a form of deviance and in which the unsuccessful are viewed by many as being immoral, in science failure to achieve eminence does not bring moral disapprobation. It is possible that failure in a system that is accepted as equitable and in which all are believed to contribute to common goals is not as difficult to accept as failure in a system which is deemed inequitable and in which failure is morally disapproved.

Thus far we have attempted to explain the acceptance of failure in terms of the ideological or normative structure of science. There is some new evidence, however, suggesting that the status structure of science may also act to reduce the perception of failure. Hargens has recently presented data on status mobility among American

---

career. Among the early producers there does not seem to be any decline in the quality of later work.

academic scientists that point to a dual process of status changes among scientists over time.[39] He examined the mobility of 1,700 academic scientists between 1961 and 1966. Using the Cartter rankings, he divided graduate institutions into six prestige-levels and looked at the movement of individuals between the levels over time. He also collected data on the changing academic rank of the scientists. His data are instructive for they indicate that a high proportion of those scientists who moved from higher- to lower-prestige departments almost invariably had received promotions in rank. Eighty percent of the mobile scientists who were downwardly mobile were upwardly mobile in rank, compared to 53 percent of the scientists who were upwardly mobile in prestige of academic affiliation. This pattern was true for scientists at all academic ranks. Hargens suggests that this dual feature of the status system of science tends to minimize perceptions of failure among scientists.

Finally, it must be made clear that when we use the term "failure" it is only in a relative sense. As we pointed out in chapter 3, all scientific occupations are accorded high prestige by the society at large. Thus, a solid-state physicist who has published a few papers that have received no recognition is still a "physicist" and, as such, will be respected by his nonphysicist friends, neighbors, and family. It is satisfying to be a member of a high-prestige occupation even if one is not greatly successful in that occupation. Furthermore, we have been considering success in the national scientific community.

At some point, usually early in his career, the average physicist realizes that he is not going to win the Nobel Prize and that he even has very little chance of getting a tenure position at a major department. He will then turn his attention to achieving success in what might be called the "minor leagues." We hypothesize here that physicists who are not successful in achieving national recognition are likely to adopt local people as reference groups and drop the national scientific elite as a meaningful reference group. Physicists in the "minor leagues" will spend their time teaching, doing ad-

39. Lowell L. Hargens, "Status Mobility of American Academic Scientists" (Paper presented at the annual meeting of the American Association for the Advancement of Science, Chicago, December 28, 1970). We thank Professor Hargens for allowing us to cite his findings before their publication. Hargens does not in his paper link this process with the actual perceptions of scientists of their "success" or "failure." Thus it remains conjectural as to whether this process actually does affect perceptions of failure in science.

ministrative work, and even doing a small amount of research for the fun of it. Many of these physicists probably achieve a statisfying degree of recognition in their local environments. Local prestige probably goes a long way to make up for failure to achieve national recognition.

Appendix A
DESCRIPTION OF SAMPLES AND DATA

One significant feature of the data reported in this book is that they are not all generated from a single sample of American scientists. Rather, the data reported in these pages come from a series of studies that focus on the problem of stratification in science. The variety of samples of scientists that were used to study these various problems were drawn because of their appropriateness for dealing with a particular topic. Thus, a study of the reward system of modern physics called for several samples of physicists. We had to represent adequately the population of university physicists, thus a random sample of university physicists was generated through the use of a short questionnaire mailed to some 2,000 physicists at 86 physics departments that have offered doctorate degrees for an extended period of time (see full discussion below of sample of 1,308 physicists). We also wanted, however, to examine in detail the behavior of elite physicists. Therefore, we had to oversample eminent scientists, since the social system of physics is so stratified that a random sample would not adequately identify a sufficient number of truly eminent scientists. Consequently, we generated a relatively small sample of 120 university physicists, selected on the basis of four stratifying criteria that overrepresented eminent scientists (see full discussion of the sample of 120 physicists below). But take an even more extreme example of the need for generating different samples for different purposes: the case of ascertaining the extent of discrimination against women scientists. A random sample of physicists would turn up virtually no women. In fact, this was found to be the case for our random sample of 1,308 university physicists. In fact, to examine the treatment that women receive in science by looking at physics would itself be an error, since women comprise only some 2 percent of the population of American physicists. Consequently, it was necessary to grossly oversample women scientists if we wanted to adequately assess their treatment. Consequently, we had to seek out yet another data set, one which would provide as much information on as many women scientists as possible, and one which covered a number of fields of science.

There are another set of reasons for the unusually large number of samples that are presented in this book. The aim of this research effort was to explore various interesting questions in the sociology of science in order to begin to develop this relatively new sociological specialty. This, in fact, was the explicit goal of the program grant that we received from the National Science Foundation. When we started these inquiries some five years ago, we were not compelled to follow any preordained line of analysis, nor did we have to set to work at answering specific questions contained in a proposal. We were to follow the lines of inquiry that we believed led to the most fruitful results in developing a knowledge of the social aspects of science. Within the framework of such a program structure, many of the paths that we have taken were not envisioned when we started on this project. In the process of working on a particular problem which yielded certain answers to the questions that we posed, other interesting questions emerged that were thematically linked but which called for the collection of additional data. Additional data collection inevitably called for sampling for slightly new study designs, which in turn yielded new samples or subsamples that were more appropriate for the questions that we were now addressing.

In some sense the varieties of samples that were used in the analysis in these pages do not present the most elegant form of data collection. If we had the advantages of hindsight we might be able to reduce the number of samples that we actually used. But the pattern of investigation that is used to explore unexplored ground in a discipline often does not yield elegant designs for which a single sample can be generated. In order to help identify the samples and subsamples used in this book, we provide below a brief description of some of the samples which are referred to in the monograph.

THE PRIMARY SAMPLES

Two primary samples of scientists provide the basis for much of the analysis in this book. In this Appendix we will briefly describe how these samples were generated. We will also briefly describe data contained in several other samples which are widely used in the monograph.

1. *The sample of 120 physicists.* This sample was selected exclusively from the population of American university physicists. But the sample is actually more exclusive than this. It was drawn from the faculties of physics departments that had granted at least one Ph.D. annually during the period 1952–62. In this respect, the sample paralleled the criteria of inclusion used by Allan Cartter in his study, *An Assessment of the Quality of Graduate Education in the United States*. A total of 86 university departments were included in the universe of the study. Within each of these departments we took every member of the physics department who was listed in the bulletin of the university, and included him in our population. The source for these data was: *Doctoral Programs in Physics: A Handbook for Advisors of Prospective Students* (American Institute of Physics: Pub. R-184). The population of university physicists was then stratified along four dimensions: age, prestige rank of their university department, productivity, and number of honorific awards. These data were obtained from *American Men of*

Science and *Science Abstracts.* The final sample of 120 scientists was drawn from this stratified population. We oversampled highly productive scientists, located in high-prestige departments. In fact, the sample of 120 overrepresents eminent physicists. This select sample of 120 scientists was used for a set of purposes: first, to determine the ways in which various patterns of productivity were rewarded by the social system of science; second, to gauge the factors that determine the visibility of physicists' research; third, to estimate the sources of intellectual influence on the work produced by this sample of physicists; fourth, to determine what types of scientists throughout the stratification system made use of the research produced by these scientists.

This sample does not represent a partial response to a questionnaire. Data were collected for every scientist in the selected sample. There were no substitutions of alternative cases for the sample. The sampling frame took the following form: four categories for age; three for number of honorific awards; two for productivity; and four for prestige rank of department. A total of 128 sampling cells resulted from these divisions. From this general framework the sample of 120 was randomly selected. Obviously, since we oversampled eminent scientists, there were a number of sample cells from which no scientists were selected. In most cases these were categories that represented scientists who were young, had received no awards, were not productive, and were located at less prestigious departments.

We collected a variety of data for the scientists in this sample. Productivity data were gathered from *Science Abstracts;* background characteristics and awards received were obtained from *American Men of Science;* citation frequencies came from *The Science Citation Index;* and the visibility of the work of the 120 physicists were obtained from responses by 1,308 academic physicists to a questionnaire that asked them to indicate their familiarity with the work of the select sample of 120. We turn now to a brief description of several subsamples that were generated from the basic data set of 120 physicists.

1.1 *Sample of 84 and 385 university physicists.* These samples are primarily used in chapter 8 where we discuss patterns of influence on scientific discoveries. The sample of 84 physicists excludes all scientists in the sample of 120 who had either not received any citations to their work in the 1965 *SCI* or who had only published review articles. Thus we had data on the social characteristics of the 84 physicists and their publication and citation records. From the work of these 84 physicists we generated a sample of 385 authors cited in the "best" paper produced by each of the 84 source physicists, as defined by the number of citations that the paper received in the 1965 *SCI.* This sample of 385 scientists represented a one-third random sample of all authors cited in the best papers of the 84 physicists. For this group of 385 cited authors, we collected background data on age, specialty, current affiliation, source of doctorate, number of honorific awards, publications, and citations. We did not have data on their visibility. All of the background information was obtained from the *AMS* and the various science abstracts.

2. *The sample of 1,308 physicists.* This sample of university physicists was obtained from returns of a short questionnaire sent to the members of academic physics departments that had granted at least one Ph.D. annually during the period 1952–62. Criteria for inclusion in the population were once again those used by Cartter. Eighty-six departments of physics were included in the universe; a total of 2,036 physicists received a copy of the questionnaire. Responses were obtained from a total of 1,360 physicists, of which 1,308, or 66.8%, were useable. Most of the questionnaires that could not be used were incomplete (that is, they were only partially filled out in terms of requested information, or the physicist chose not to give us his own name). There were few identifiable sampling biases. Physicists in highly-ranked departments responded to the questionnaire in the same proportions as those in lower-ranked departments; those with tenure rank to the same extent as those without tenure. So far as we can tell, the 1,308 physicists returning questionnaires are representative of physicists in American universities.

Responses by the 1,308 physicists were used to compute both visibility and awareness scores for the 120 physicists. Thus, each of these 1,308 physicists acted as a judge of the visibility of other physicists' work. These physicists also judged the prestige of a sample of 98 awards that were assembled from *American Men of Science* and *Physics Today*. These awards represented a large sample of awards granted to physicists. Since it was not feasible to ask each of the university physicists to rank all 98 awards, we used five different forms of the questionnaire. Ten awards were included on all five forms. Since the difference between the scores of these awards which appeared on all five forms was statistically insignificant, we conclude that the score received by each award is representative of its prestige among academic physicists. As an example of the closeness of the ratings on the five forms, membership in the National Academy of Science received ratings of 4.28, 4.32, 4.03, 4.24, and 4.27 on the five forms. The range of prestige scores was from 5.0 (highest) to 1.0 (lowest). For the exact wording of the question see Appendix B. For a discussion of variations in the response to a set of specific awards, in which the population of physicists were given a different set of identifying clues, see above, pp. 184–87.

A similar multiform procedure was used to achieve rankings of the visibility of the 120 physicists. Twenty-five physicists' names appeared on each of the five forms of the questionnaire. On each of the forms one name was fictitious, so that we could test the level of guessing among the population of judges. It turned out that the level of guessing did not exceed 5 percent; it was no greater for a fictitious Chinese name than an American name. There were no systematic differences in patterns of guessing—that is, there were no differences in the guessing rate among full professors as against lower-ranked professors; eminent and noneminent physicists, and so on.

2.1 *Samples of 171 and 861 scientists.* These samples were used in chapter 2 to determine whether weighting the number of citations that a cited author received by the quality of the work produced by the men who cited his work, would lead to any different substantive results than a straight count of citations. The 171 scientists were drawn from the sample of 1,308 physicists. We stratified the 1,308 university physicists along three dimensions: specialty, assessed rank of university department, and quality of scientific research. From within each stratum a random sample was chosen, totalling 171 physicists. For each of these 171 physicists we selected a sample of 15 or fewer users or citers of their work as listed in the 1965 *SCI.* Since some of the cited authors had fewer than 15 citations, we took all of the citers of these authors. Social and individual characteristics for both of these samples were collected from *AMS, Science Abstracts,* and the *SCI.*

2.2 *Sample of 91 university physicists.* This sample, used in chapter 7, was drawn from the sample of 1,308 university physicists in the following way: we identified all those university physicists who had published a paper before 1961 which had received at least 10 citations in the year 1966. There were 91 physicists who met this standard. The sample represents a subgroup of physicists who published highly significant papers as measured by citations at least five years after the publication of that paper. For this group of physicists we not only had all of the data on stratification variables that we had for the 1,308, but we also collected additional data on citations to this group: primarily in the form of citations to individual papers in different years as well as citations to the entire life work of the scientist.

2.3 *Sample of 157 physicists.* This too was used to study the Matthew Effect in chapter 7. It was a subsample of the 1,308 physicists drawn by the criteria of including all scientists whose life work published before 1961 received 20 or more citations in the 1966 *SCI.* The stratification information for this sample is identical to that obtained for the 1,308 physicists, since this was nothing more than a subsample of "elite" scientists in terms of the quality of their publications.

3. *Sample of 300 full professors in five scientific fields.* The sample consists of 60 full professors at Ph.D.-granting institutions in each of the following five fields: physics, chemistry, biochemistry, psychology, and sociology. These data were originally collected for a comparative study of scientific reward systems (See Stephen Cole, "Scientific Reward Systems"). Every member of the sample was promoted from associate to full professor between 1965 and 1969. All scientists meeting this criterion and teaching in any of the departments listed in the 1968 American Council on Education survey of graduate education were listed (see K. D. Roose and Charles J. Anderson, *A Rating of Graduate Programs* [Washington, D.C.: American Council on Education, 1970]). We then randomly selected

20 scientists from each of the high-, medium-, and low-ranked groups of departments in each field. The names of the physicists were taken from the 1969 American Physical Society's *Directory of Physics and Astronomy Faculties;* chemists from the American Chemical Society's *Directory of Members, 1967;* psychologists from the American Psychological Association's *Directory of Members, 1970;* and sociologists from the American Sociological Association's *Guide to Graduate Study, 1970.* It was very difficult to get a list of all scientists working in biochemistry. We used the *Directory of Members, 1970,* of the Federation of American Societies for Experimental Biology. Information on career history, age, and awards received for each scientist was collected from *American Men of Science* and the directories of the respective professional associations. The quality of published work was measured by the number of citations listed for each scientist in the 1965 and 1969 *SCI.* (Citations to sociologists were counted separately.) Information on the number of publications was collected from the appropriate abstracting journals, except for sociology, where the information was obtained from the vitae of the scientists. Since the distribution on many of the variables differed from field to field, it was necessary to standardize variables separately for each field. All results presented are based on these standardized scores.

4. *Sample of 499 and 754 men and women scientists.* The sample of 499 men and women scientists is used in chapter 5, where we discuss the influence of sex status on scientific recognition. The sample was generated in the following way. In 1969 Dr. Helen Astin published a study of women doctorates in the United States. Her study, which focused exclusively on women in all academic fields who received their doctorates in 1957 and 1958, was based upon a returned mail questionnaire from slightly more than 1,500 women. The survey was administered in 1965, some seven to eight years after the receipt of the doctorate. Dr. Astin gave us permission to use her data set for secondary analysis and to generate a matched sample of male doctorates. This we did by matching every woman who had received her doctorate in one of the physical, biological, or social sciences (including history) with a male doctorate who received his degree from the same university, in the same year, in the same field, and in the same specialty. These male matches were drawn from the doctorate record file of the Office of Scientific Personnel. All names were kept completely confidential. We never had access to both names of scientists and data on them at the same time. There were 764 women doctorates who needed matches. Using our sampling procedure, we could find matches for only 749 of the women. A pool of available male matches for each female was generated, and two men were selected randomly as matches for each female. Only one of these was used. We generated two for each woman in order to raise the probability of finding one of them in *American Men of Science* (as of the twelfth edition, *American Men and Women of Science*). We had data on the high school records, I.Q.'s, source of graduate training, for most

of the men and women from the Astin sample and the OSP file. We collected additional data on the publication records of the men and women from the various publication abstracts in various fields; the source index of *SCI;* citation counts from the *SCI;* and background data on marital and family status, and postdoctorate experience from *AMS*. Data on publications and citations were not readily available for doctorates in most of the social sciences and in the humanities. Many of the men and women scientists had not followed academic career lines. Since nonacademic careers involve highly disparate reward and stratification systems, we decided for the purposes of this particular study to concentrate only on academic scientists who were employed full time as of 1965 within the physical, biological, and social sciences and only on those fields for which we had role-performance data. From this group of fields we excluded physics because there were virtually no women doctorates in 1957 and 1958. We excluded all social sciences other than psychology because we had complete publication and citation data only for psychologists. This left us with a total of 499 academic men and women scientists in chemistry, biological sciences (of various specialties), and psychology. The analysis presented in chapter 5 is based on these 499 cases. All of the evidence that we have suggests that even though we deal with a subsample for which we generated matches, the matching element in the study design has not been lost. For example, there remains virtually no correlation between sex status and rank of doctorate department. We did not have complete data for all variables on the 499 cases. For example, some women could not be found in *AMS,* and consequently some later career data for them are missing. Consequently, the correlation and regression analysis on the sample is based upon a "pair-wise" deletion of missing values.

The sample of 754 men and women scientists includes all men and women in chemistry, biological sciences, and psychology, regardless of whether they held academic jobs. The sample is used only in chapter 2 for the purpose of examining the stability of citations over an extended period of time. It was the one large sample of scientists for which we had citation data at six points in time: 1961, 1964, 1965, 1967, 1969, 1970. All of these citation counts were a straight count of the total number of citations following a scientist's name in the *SCI*. All self-citations were, as usual, not included in the counts. The pattern of findings for stability of citations were the same when the data were analyzed for men scientists alone.

Appendix B DATA ON HONORIFIC AWARDS

Visibility and Prestige Scores for Sample of 98 Honorific Awards and Postdoctoral Fellowships Held by Physicists

	Name of Award	Visibility of Awards		Prestige of Awards	
		Score *	Form No.	Score †	(N)
1	Nobel Prize	100	I–V	4.98	1277
2	Member National Academy of Sciences	95	I–V	4.22	1261
3	Fulbright Scholar or Lecturer	94	III	2.58	236
4	National Science Foundation Fellow	93	IV	2.43	248
5	Guggenheim Fellow	92	I–V	3.14	1170
6	Enrico Fermi Award	92	I–V	4.31	1170
7	Rhodes Scholar	91	III	3.2	236
8	Member, Royal Society	86	III	4.01	216
9	Sloan Foundation Fellow	83	I–V	3.18	1061
10	Honorary Degree, Harvard	81	III	3.7	208
11	Oersted Medal (American Association of Physics Teachers)	80	III	3.31	190
12	Honorary Degree, Berkeley	78	IV	3.1	200
13	Ford Foundation Fellow	72	I	2.69	186
14	Oliver E. Buckley Prize (American Association of Physics Teachers)	66	III	3.65	184
15	National Research Council Fellow	68	I–IV	2.97	688
16	Dannie Heineman Prize (American Institute of Physics)	66	III	3.8	171
17	Honorary Degree, Johns Hopkins	65	I	3.00	168

	Name of Award	Visibility of Awards Score *	Form No.	Prestige of Awards Score †	(N)
18	Fritz London Award	65	V	4.03	169
19	Honorary Degree, University of North Carolina	64	V	2.2	168
20	Rockefeller Foundation Fellow	61	IV	2.9	161
21	Ernest Orlando Lawrence (Atomic Energy Commission) Award	57	IV	3.8	163
22	Atomic Energy Commission Fellow	59	II	2.3	144
23	Member, Academie Français	56	I	4.1	144
24	Presidential Award for Distinguished Civilian Service	56	III	3.4	139
25	American Physical Society Prize	54	IV	3.4	143
26	Presidential Medal for Merit	53	II	3.7	129
27	Atoms for Peace Prize	51	IV	3.8	135
28	National Institutes of Health Fellowship	49	III	1.8	112
29	Honorary Degree, Vanderbilt University	49	II	2.3	127
30	Westinghouse Research Fellow	49	I	2.1	127
31	Karl Taylor Compton Gold Medal (American Institute of Physics)	48	IV	3.7	127
32	Albert Einstein Gold Medal	47	I	4.2	122
33	Max Planck Medal (Physical Society of London)	47	II	4.0	120
34	Carnegie Fellow	43	I	2.5	110
35	Presidential Medal of Freedom	43	I	3.9	110
36	Franklin Medal (Franklin Institute)	41	I	3.43	107
37	Rumford Medal (Premium) (American Academy of Arts and Sciences)	43	III	3.27	98
38	National Medal of Science	38	IV	4.02	88
39	Research Corporation Award	38	II	2.59	93
40	Lorentz Medal (Royal Netherlands Academy of Science)	36	II	4.0	87
41	Peter Debye Award (American Chemical Society)	41	IV	3.8	81

	Name of Award	Visibility of Awards Score *	Form No.	Prestige of Awards Score †	(N)
42	Langmuir Award (American Chemical Society)	30	III	3.77	75
43	Priestley Medal	30	V	3.74	79
44	Royal Astronomical Society Gold Medal	30	II	4.2	84
45	Henry Draper Medal (National Academy of Sciences)	28	III	3.68	71
46	U.S. Public Health Service Fellow	28	II	2.0	68
47	Frederick Ives Medal (Optical Society of America)	23	I–V	3.42	287
48	American Philosophical Society Fellow	20	II	2.79	47
49	Comstock Prize (National Academy of Sciences)	16	V	3.5	40
50	Medal of Honor (Institute of Radio Engineers)	16	I	2.8	42
51	U.S. Department of Commerce Medal	15	I	2.8	38
52	Elliot Medal (National Academy of Sciences)	14	I	3.8	37
53	Jewitt Fellow	14	V	3.1	236
54	Longstreth Medal (Franklin Institute)	14	II	2.5	35
55	Ballantine Medal (Franklin Institute)	14	II	2.6	36
56	Adolf Lomb Medal (Optical Society)	14	I	3.1	35
57	U.S. Navy Meritorious Civilian Service Award	14	II	3.5	37
58	Proctor and Gamble Fellow	13	III	2.2	34
59	Scientific Research Society Pure Science Award	13	III	2.4	33
60	Cressy Morrison Prize (New York Academy of Science)	11	III	2.5	29
61	Eugene C. Bingham Medal (Society of Rheology)	10	III	2.8	27
62	Gilbert Lewis Medal (Chemical Society)	10	II	3.5	26
63	Kalinga Prize (UNESCO)	10	IV	3.1	27

	Name of Award	Visibility of Awards Score *	Form No.	Prestige of Awards Score †	(N)
64	Mark Mills Award (American Nuclear Society)	9	IV	2.3	24
65	National Education Board Fellow	9	V	2.2	33
66	Newcomb Cleveland Prize (American Association for the Advancement of Science)	8	IV	2.9	21
67	Duddell Medal and Prize (British Institute of Physics)	8	IV	3.4	22
68	Louis Levy Medal (Franklin Institute)	8	I	2.8	20
69	Elliot Cresson Medal (Franklin Institute)	8	I–V	3.1	107
70	Benjamin Lamme Medal (Journal of Engineering Education)	8	I	2.3	22
71	Henry Fellow	7	II	2.5	18
72	Arthur S. Fleming Award (American Chemical Society)	6	IV	3.7	17
73	William Nicholas Award (American Chemical Society)	6	II	3.5	15
74	Charles P. Walcott Medal (National Academy of Sciences)	6	IV	2.9	15
75	J. Lawrence Smith Medal (National Academy of Sciences)	6	IV	3.5	17
76	Morris Lieberman Award (Institute of Radio Engineers)	6	II	2.6	17
77	Nuclear Pioneer Award	6	I	2.8	16
78	Edgar D. Tillyer Medal (Optical Society)	6	II	3.0	16
79	Bruce Gold Medal (Astronomical Society of the Pacific)	6	I	3.2	17
80	Barnard Medal (Columbia University)	6	I	2.8	16
81	John Scott Medal (City of Philadelphia)	5	II	2.7	12
82	Dudley Medal (American Society of Testing and Materials)	4	I–V	2.3	49
83	William Bowie Medal (American Geophysical Union)	4	IV	3.1	11

	Name of Award	Visibility of Awards Score *	Form No.	Prestige of Awards Score †	(N)
84	Canizzaro Gold Medal (Italian National Academy)	4	IV	3.3	10
85	Capt. Robert Dexter Conrad Award (Office of Naval Research)	4	I–V	2.7	46
86	Emile Berliner Award (Audio Engineering Society)	3	I	2.12	8
87	Pregel Prize (New York Academy of Science)	3	III	2.0	8
88	Lloyd Fellow	3	IV	1.9	7
89	Lawrence Langwald Award (Lawrence C. Langwald Foundation)	3	II	2.9	7
90	Day Medal	2	V	3.0	5
91	Vetlesen Prize (Columbia University)	2	I	3.2	5
92	General Donovan Memorial Award	2	II	2.7	4
93	Carty Medal	2	V	2.2	5
94	Douglas Medal (American Society for Testing and Materials)	2	II	2.2	5
95	Pupin Gold Medal ‡	18	V	3.7	48
96	Arthur Trachtenberg Prize ‡	5	I	2.4	13
97	Richard Saunders Gold Medal ‡	2	IV	2.6	5
98	Norwood Prize ‡	2	III	1.6	5

Total Number of Useable Responses to Each of the Five Forms of the Awards Questionnaire:

Form I	258
Form II	245
Form III	250
Form IV	264
Form V	261
Total	1,278

*The visibility of awards was determined by the proportion of the total number of physicists who gave a rating from one to six to an award. The actual score given was not taken into account when computing "visibility scores." The base figures on which percentages were computed differed from one form to another. The forms on which the award appeared are listed next to the visibility score.

†The range of possible scores for prestige was 5 (highest) to 1 (lowest). As noted above, the prestige score was the mean score of those physicists who rated the award. Prestige scores thus excluded the responses of those physicists who never heard of the award or who did not have enough information to judge its prestige. The exact wording of the question was:

"The following thirty awards represent a sample of several kinds of awards. For those which are known to you, we ask that you indicate your judgment of its prestige by circling one of the five rankings. You may not have heard of many of these awards, as most are not widely known. If you have heard of an award but do not know enough about it to evaluate its prestige, please circle No. 6. If you have never heard of the award, circle No. 7. The circling of either 6 or 7 provides useful information as a ranking since it will indicate which awards are least known among physicists."

Name of Award	*Rough Estimate of Prestige* *High Prestige*				*Low Prestige*	*Have heard of award but don't have enough information to judge its prestige*	*Have never heard of the award*
1. Nobel Prize	1	2	3	4	5	6	7

(When we computed the prestige scores we inverted the values, so that the higher the value an award was given, the higher its score would be.)

‡These are fictional awards, used to measure exaggeration of knowledge.

NAME INDEX

SUBJECT INDEX